Optics for AI and AI for Optics

Optics for AI and AI for Optics

Special Issue Editors

Jinlong Wei
Alan Pak Tao Lau
Lilin Yi
Elias Giacoumidis
Qixiang Cheng

MDPI • Basel • Beijing • Wuhan • Barcelona • Belgrade • Manchester • Tokyo • Cluj • Tianjin

MDPI

Special Issue Editors
Jinlong Wei
Huawei Technologies Düsseldorf GmbH
Germany

Lilin Yi
Shanghai Jiao Tong University
China

Elias Giacoumidis
VPIphotonics GmbH
Germany

Alan Pak Tao Lau
Hong Kong Polytechnic University
Hong Kong

Qixiang Cheng
University of Cambridge
UK

Editorial Office
MDPI
St. Alban-Anlage 66
4052 Basel, Switzerland

This is a reprint of articles from the Special Issue published online in the open access journal *Applied Sciences* (ISSN 2076-3417) (available at: https://www.mdpi.com/journal/applsci/special_issues/optics_AI).

For citation purposes, cite each article independently as indicated on the article page online and as indicated below:

LastName, A.A.; LastName, B.B.; LastName, C.C. Article Title. *Journal Name* **Year**, *Article Number*, Page Range.

ISBN 978-3-03936-398-8 (Hbk)
ISBN 978-3-03936-399-5 (PDF)

Contents

About the Special Issue Editors vii

Jinlong Wei, Lilin Yi, Elias Giacoumidis, Qixiang Cheng and Alan Pak Tao Lau
Special Issue on "Optics for AI and AI for Optics"
Reprinted from: *Appl. Sci.* **2020**, *10*, 3262, doi:10.3390/app10093262 1

Bin Shi, Nicola Calabretta and Ripalta Stabile
Numerical Simulation of an InP Photonic Integrated Cross-Connect for Deep Neural Networks on Chip
Reprinted from: *Appl. Sci.* **2020**, *10*, 474, doi:10.3390/app10020474 5

Chen Chen, Xiong Deng, Yanbing Yang, Pengfei Du, Helin Yang and Lifan Zhao
LED Nonlinearity Estimation and Compensation in VLC Systems Using Probabilistic Bayesian Learning
Reprinted from: *Appl. Sci.* **2019**, *9*, 2711, doi:10.3390/app9132711 21

Maximilian Schaedler, Maxim Kuschnerov, Fabio Pittalà, Stefano Calabrò, Christian Bluemm and Stephan Pachnicke
Deep Neural Network Equalization for Optical Short Reach Communication
Reprinted from: *Appl. Sci.* **2019**, *9*, 4675, doi:10.3390/app9214675 33

Stenio M. Ranzini, Francesco Da Ros, Henning Bülow and Darko Zibar
Tunable Optoelectronic Chromatic Dispersion Compensation Based on Machine Learning for Short-Reach Transmission
Reprinted from: *Appl. Sci.* **2019**, *9*, 4332, doi:10.3390/app9204332 47

Haide Wang, Ji Zhou, Yizhao Wang, Jinlong Wei, Weiping Liu, Changyuan Yu and Zhaohui Li
Optimization Algorithms of Neural Networks for Traditional Time-Domain Equalizer in Optical Communications
Reprinted from: *Appl. Sci.* **2019**, *9*, 3907, doi:10.3390/app9183907 57

Rebekka Weixer, Jonas Koch, Patrick Plany and Stephan Pachnicke
Mitigation of Nonlinear Impairments by Using Support Vector Machine and Nonlinear Volterra Equalizer
Reprinted from: *Appl. Sci.* **2019**, *9*, 3800, doi:10.3390/app9183800 67

Ivan Aldaya, Elias Giacoumidis, Geraldo de Oliveira, Jinlong Wei, Julián Leonel Pita, Jorge Diego Marconi, Eric Alberto Mello Fagotto, Liam Barry and Marcelo Luis Francisco Abbade
Histogram Based Clustering for Nonlinear Compensation in Long Reach Coherent Passive Optical Networks
Reprinted from: *Appl. Sci.* **2020**, *10*, 152, doi:10.3390/app10010152 77

Elias Giacoumidis, Yi Lin, Mutsam Jarajreh, Sean O'Duill, Kevin McGuinness, Paul F. Whelan and Liam P. Barry
A Blind Nonlinearity Compensator Using DBSCAN Clustering for Coherent Optical Transmission Systems
Reprinted from: *Appl. Sci.* **2019**, *9*, 4398, doi:10.3390/app9204398 91

Mutsam A. Jarajreh
Reduced-Complexity Artificial Neural Network Equalization for Ultra-High-Spectral-Efficient Optical Fast-OFDM Signals
Reprinted from: *Appl. Sci.* **2019**, *9*, 4038, doi:10.3390/app9194038 . . . . **99**

Xiaomin Liu, Huazhi Lun, Mengfan Fu, Yunyun Fan, Lilin Yi, Weisheng Hu and Qunbi Zhuge
AI-Based Modeling and Monitoring Techniques for Future Intelligent Elastic Optical Networks
Reprinted from: *Appl. Sci.* **2020**, *10*, 363, doi:10.3390/app10010363 . . . . **111**

Qianwu Zhang, Hai Zhou, Yuntong Jiang, Bingyao Cao, Yingchun Li, Yingxiong Song, Jian Chen, Junjie Zhang and Min Wang
A Simple Joint Modulation Format Identification and OSNR Monitoring Scheme for IMDD OOFDM Transceivers Using K-Nearest Neighbor Algorithm
Reprinted from: *Appl. Sci.* **2019**, *9*, 3892, doi:10.3390/app9183892 . . . . **129**

Panagiotis Giounanlis
Photon Enhanced Interaction and Entanglement in Semiconductor Position-Based Qubits
Reprinted from: *Appl. Sci.* **2019**, *9*, 4534, doi:10.3390/app9214534 . . . . **141**

About the Special Issue Editors

Jinlong Wei received his Ph.D. degree in Electronic Engineering from the Bangor University, Bangor, UK, in 2011. After receiving his Ph.D., he joined the University of Cambridge, UK, as a research associate (2011–2014). He was awarded an EU Marie Curie fellowship and conducted the award research at AVDA Optical Networking SE, Germany (2014–2016). In 2016, he became a senior researcher at Huawei Technologies German Research Center, Munich, Germany, where he is currently a principal researcher. His research interests include modulations, (intelligent) signal processing, and devices for high-speed optical communication systems and networks. His various pioneering works on optical access and data center networks were reported by BBC, Reuters, Yahoo, OSA, etc. He is a senior member of IEEE, a Marie Curie Fellow, and an honorary research fellow of Bangor University.

Alan Pak Tao Lau received his B.A.Sc. degree in Engineering Science (Electrical Engineering option) and his M.A.Sc. degree in Electrical and Computer Engineering from the University of Toronto, Toronto, ON, Canada, in 2003 and 2004, respectively. He received his Ph.D. degree in Electrical Engineering from Stanford University, Stanford, CA, USA, in 2008. In 2008, he became an assistant professor at the Hong Kong Polytechnic University, where he is currently a professor. He collaborates with industry in various aspects of optical communications and serves in organizing committees of numerous conferences in optical communications. His current research interests include long-haul and short-reach coherent optical communication systems, optical performance monitoring, and machine learning applications in optical communications and networks.

Lilin Yi received his Ph.D. degree from the Ecole Nationale Supérieure des Télécommunications (ENST, currently named Telecom ParisTech), France, and Shanghai Jiao Tong University, China, in March and June 2008, respectively, as a joint-educated Ph.D. student. He is currently a full professor at Shanghai Jiao Tong University. His main research topics include high-speed optical communications, intelligent mode-locking fiber lasers, optical signal processing, and machine-learning-based digital signal processing. Dr. Lilin Yi is the author or co-author of more than 180 papers in peer-reviewed journals and conferences, including invited papers/invited talks in JLT/OFC/ECOC. Dr. Yi has earned the "Young Scholars of the Yangtze River in China" and "National Science Fund for Excellent Young Scholars of China" awards. He serves as a TPC member of OFC/OECC/CLEO-PR/ACP and as TPC track/workshop/symposium co-chair of OFC/ECOC/OECC/CLEO-PR/ACP. He is an associate editor of *Optical Fiber Technology*.

Elias Giacoumidis received his Ph.D. in 2011 from Bangor University of Wales (UK). He is currently working at VPIphotonics (Berlin, Germany) as a project manager. He was previously a lecturer in electronic engineering at Beijing-Dublin International College of University College Dublin (2019) and a Marie Curie research fellow in optical communications at Dublin City University and SFI CONNECT Research Centre of Ireland (2017–2019). He was a collaborative researcher at Xilinx, Ireland (2018), where he developed the world's first real-time machine-learning-based nonlinearity compensator for high-speed fiber-optic networks. From 2011 to 2017, he worked for various prestigious optical communications research groups, including Heriot-Watt University, University of Sydney, Aston University, Telecom-ParisTech, and Athens Information Technology. Dr. Giacoumidis is a member of IEEE and OSA and was nominated as outstanding reviewer of 2016 for the IEEE/OSA *Journal of Lightwave Technology*. He was nominated for the best application of AI in an academic research body (Irish AI Awards, 2019).

Qixiang Cheng received his B.S. degree from the Huazhong University of Science and Technology, Wuhan, China, in 2010 and his Ph.D. degree from the University of Cambridge, Cambridge, UK, in 2014. He then joined the Huawei Shannon Laboratory, where he researched future optical computing systems. From September 2016 to November 2019, he was first a postdoctoral researcher and then a research scientist with the Lightwave Research Lab, Columbia University, New York, NY, USA. In January 2020, he was appointed as a university lecturer in photonic devices and systems at the University of Cambridge, UK. His current research interests focus on system-wide photonic integrated circuits for optical communication and computing applications, including a range of optical functional circuits such as packet-, circuit-, and wavelength-level optical switch fabrics; massively parallel transceivers; optical neural networks; and optical networks-on-chips.

applied sciences

MDPI

Editorial

Special Issue on "Optics for AI and AI for Optics"

Jinlong Wei [1,*], Lilin Yi [2], Elias Giacoumidis [3], Qixiang Cheng [4] and Alan Pak Tao Lau [5]

1 Optical and Quantum Communication Laboratory, German Research Center, Huawei Technologies Düsseldorf GmbH, 80992 Munich, Germany
2 Department of Electronic Engineering, Shanghai Jiao Tong University, Shanghai 200240, China; lilinyi@sjtu.edu.cn
3 VPIphotonics GmbH, 10587 Berlin, Germany; ilias.giakoumidis@vpiphotonics.com
4 Electrical Engineering Division, Department of Engineering, University of Cambridge, Cambridge CB3 0FA, UK; qc223@cam.ac.uk
5 Photonics Research Center, Department of Electrical Engineering, The Hong Kong Polytechnic University, Hung Hom, Kowloon, Hong Kong SAR, China; eeaptlau@polyu.edu.hk
* Correspondence: jinlongwei2@huawei.com; Tel.: +49-89-1588344227

Received: 12 April 2020; Accepted: 27 April 2020; Published: 8 May 2020

We live in an era of information explosion and digital revolution that has resulted in rapid technological developments in different aspects of life. Artificial intelligence (AI) is playing an increasingly important role in this digital transformation. AI applications require edge cloud computing with low latency connections, where the significant challenge is that it needs a lot of computer processing power. Recently, the implementation of AI based on optics hardware [1–5] has become a popular topic due to its fundamentally lower power consumption and faster computation.

On the other hand, as the underlying basis of modern tele- and data-communications, optical networking becomes more and more complex, driven by more data and more connections. Generating, transmitting, and recovering such high-volume data requires advanced signal processing and networking technologies with high performance and cost-and-power efficiency. AI is especially useful for optimization and performance prediction for systems that exhibit complex behaviors [6–20]. In this aspect, traditional signal processing algorithms may not be as efficient as AI algorithms. AI methods have recently entered the field of optics, ranging from quantum mechanics to nanophotonic, optical communication, and optical networks.

The Special Issue is launched to bring optics and AI together to address the challenges that each face, which are difficult to address alone. There are 12 selected contributions for the special session, representing the fascinating progress in the combined area of optics and AI, ranging from photonic neural network (NN) architecture [5] to AI-enabled advances in optical communications including both physical layer transceiver signal processing [10–17] and network layer performance monitoring [18,19], as well as the potential role of AI in quantum communications [20].

Photonic neural network architecture: Bin Shi and co-workers proposed a novel photonic accelerator architecture based on a broadcast-and-weight approach for a deep NN through a photonic integrated cross-connect [5]. A three-layer NN for image classification was tested and it shows that each photonic neural layer can achieve an accuracy higher than 85%. It offers insights for the design of scalable photonic NNs to a higher dimension for solving higher complexity problems.

The applications of AI, especially machine learning in the field of optical communications, are more popular as reflected in the book. At the physical transceiver layer, the most discussed topic is the use of machine learning for various linear and nonlinear effects mitigation in optical communication systems ranging from short-reach to long-haul applications.

AI for short-reach optical communications: For short visible light communications, Chen Chen et al. introduced a probabilistic Bayesian learning algorithm to compensate the light-emitting diode (LED) nonlinearity [10]. Maximilian Schaedler and his colleagues investigated a deep NN-based nonlinear equalizer in a single lambda 600Gbps coherent short-reach link and show its superior performance compared with the conventional Volterra nonlinear equalizer [11]. Stenio M. Ranzini and co-workers focused on machine learning-aided tunable chromatic dispersion compensation using a hybrid optical and digital structure in a high-speed short-reach optical link [12]. Specifically, Haide Wang and collaborators presented an interesting work, where the NN itself was not used, but its widely used optimization approaches including the batch gradient descent (BGD) method, adaptive gradients (AdaGrad), root mean squared propagation (RMSProp), and adaptive moment estimation (Adam) algorithms were examined and compared in a traditional gradient decent equalizer to significantly speed up and stabilize the filter tap coefficient convergence [13].

AI for medium- and long-reach optical communications: For up to 100 km single mode fiber (SMF)-based applications like data center interconnects, Rebekka Weixer et al., proposed a support vector machine-based detection of signals and its combination with the Volterra nonlinear detection, which shows the best trade-off between performance and complexity [14]. For a long-reach optical access network, Ivan Aldaya and co-workers presented a novel denominated histogram-based clustering algorithm to identify the borders of the high-density areas of the constellation and to classify the nonlinearly distorted noisy constellations [15]. For long-haul applications, one of the major issues is the fiber nonlinearity. Elias Giacoumidis et al. proposed a density-based spatial clustering of applications with noise (DBSCAN) algorithm to address this challenge, which shows a significant performance improvement compared with conventional K-means clustering [16]. In another work, Mutsam A. Jarajreh indicated that compared to the Volterra nonlinear equalizer, an NN was shown to be able to relax the requirement on other system parameters such as the signal quantization bits and clipping ratio, which is valuable for practical implementation [17].

AI for optical performance monitoring: At the network level, Xiaomin Liu and teammates presented a review which discussed the advanced optical performance monitoring enabled by AI-based modeling and prediction approaches to maximize the quality of transmission and resource utilization efficiency of elastic optical networks [18]. Then, concrete use cases are followed. Moreover, Qianwu Zhang and co-workers dedicate their work to the modulation format identification and optical signal-to-noise ratio monitoring of an optical link based on the K-nearest neighbor algorithm, which shows a similar performance, but requires less computing power compared with using the artificial NN [19].

AI for quantum communications: Finally, the book also includes an interesting work from Panagiotis Giounanlis et al. on photon entanglement [20], where how AI plays a constructive role remains a question for the interested readers to think about.

Acknowledgments: We would like to thank all authors, the many dedicated reviewers, the editor team of *Applied Science*, especially Xianyan Chen (Managing Editor) for their valuable contributions, making this Special Issue possible and successful.

Conflicts of Interest: The authors declare no conflicts of interest.

References

1. Geoffrey, W.B. A role for optics in AI hardware. *Nature* **2019**, *569*, 199–200.
2. Shen, Y.; Harris, N.C.; Skirlo, S.; Prabhu, M.; Baehr-Jones, T.; Hochberg, M.; Sun, X.; Zhao, S.; Larochelle, H.; Englund, D.; et al. Deep learning with coherent nanophotonic circuits. *Nat. Photonics* **2017**, *11*, 441–446. [CrossRef]
3. Sacha, G.M.; Varona, P. Artificial intelligence in nanotechnology. *Nanotechnology* **2013**, *24*, 452002. [CrossRef] [PubMed]
4. Cheng, Q.; Kwon, K.; Glick, M.; Bahadori, M.; Carloni, L.P.; Bergman, K. Silicon Photonics Codesign for Deep Learning. *Proc. IEEE* **2020**, *108*, 1–11. [CrossRef]
5. Shi, B.; Calabretta, N.; Jordan, R.S. Numerical Simulation of an InP Photonic Integrated Cross-Connect for Deep Neural Networks on Chip. *Appl. Sci.* **2020**, *10*, 474. [CrossRef]

6. Mata, J.; Miguel, I.; Barroso, R.J.D.; Merayo, N.; Singh, S.; Jukan, A.; Chamania, M. Artificial intelligence (AI) methods in optical networks: A comprehensive survey. *Opt. Switch. Netw.* **2018**, *28*, 43–57. [CrossRef]
7. Khan, F.N.; Fan, Q.; Lu, C.; Lau, A.P.T. An optical communication's perspective on machine learning and its applications. *IEEE J. Lightwave Technol.* **2019**, *37*, 493–516. [CrossRef]
8. Giacoumidis, E.; Lin, Y.; Wei, J.; Aldaya, I.; Tsokanos, A.; Barry, L.P. Harnessing machine learning for fiber-induced nonlinearity mitigation in long-haul coherent optical OFDM. *Future Internet* **2019**, *11*, 2. [CrossRef]
9. Pu, G.; Yi, L.; Zhang, L.; Hu, W. Intelligent programmable mode-locked fiber laser with a human-like algorithm. *Optica* **2019**, *6*, 362–369. [CrossRef]
10. Chen, C.; Deng, X.; Yang, Y.; Du, P.; Yang, H.; Zhao, L. LED Nonlinearity Estimation and Compensation in VLC Systems Using Probabilistic Bayesian Learning. *Appl. Sci.* **2019**, *9*, 2711. [CrossRef]
11. Schaedler, M.; Bluemm, C.; Kuschnerov, M.; Pittalà, F.; Calabrò, S.; Pachnicke, S. Deep Neural Network Equalization for Optical Short Reach Communication. *Appl. Sci.* **2019**, *9*, 4675. [CrossRef]
12. Ranzini, S.M.; Ros, F.D.; Bülow, H.; Zibar, D. Tunable Optoelectronic Chromatic Dispersion Compensation Based on Machine Learning for Short-Reach Transmission. *Appl. Sci.* **2019**, *9*, 4332. [CrossRef]
13. Wang, H.; Zhou, J.; Wang, Z.; Wei, J.; Liu, W.; Yu, C.; Li, Z. Optimization Algorithms of Neural Networks for Traditional Time-Domain Equalizer in Optical Communications. *Appl. Sci.* **2019**, *9*, 3907. [CrossRef]
14. Weixer, R.; Koch, J.; Plany, P.; Ohlendorf, S.; Pachnicke, S. Mitigation of Nonlinear Impairments by Using Support Vector Machine and Nonlinear Volterra Equalizer. *Appl. Sci.* **2019**, *9*, 3800. [CrossRef]
15. Aldaya, I.; Giacoumidis, E.; Oliveira, G.; Wei, J.; Pita, J.L.; Marconi, J.D.; Fagotto, E.A.M.; Barry, L.; Abbade, M.L.F. Histogram Based Clustering for Nonlinear Compensation in Long Reach Coherent Passive Optical Networks. *Appl. Sci.* **2020**, *10*, 152. [CrossRef]
16. Giacoumidis, E.; Lin, Y.; Jarajreh, M.; O'Duill, S.; McGuinness, K.; Whelan, P.F.; Barry, L.P. A Blind Nonlinearity Compensator Using DBSCAN Clustering for Coherent Optical Transmission Systems. *Appl. Sci.* **2019**, *9*, 4398. [CrossRef]
17. Jarajreh, M.A. Reduced-Complexity Artificial Neural Network Equalization for Ultra-High-Spectral-Efficient Optical Fast-OFDM Signals. *Appl. Sci.* **2019**, *9*, 4038. [CrossRef]
18. Liu, X.; Lun, H.; Fu, M.; Fan, Y.; Yi, L.; Hu, W.; Zhuge, Q. AI-Based Modeling and Monitoring Techniques for Future Intelligent Elastic Optical Networks. *Appl. Sci.* **2020**, *10*, 363. [CrossRef]
19. Zhang, Q.; Zhou, H.; Jiang, Y.; Cao, B.; Li, Y.; Song, Y.; Chen, J.; Zhang, J.; Wang, M. A Simple Joint Modulation Format Identification and OSNR Monitoring Scheme for IMDD OOFDM Transceivers Using K-Nearest Neighbor Algorithm. *Appl. Sci.* **2019**, *9*, 3892. [CrossRef]
20. Giounanlis, P.; Blokhina, E.; Leipold, D.; Staszewski, R.B. Photon Enhanced Interaction and Entanglement in Semiconductor Position-Based Qubits. *Appl. Sci.* **2019**, *9*, 4534. [CrossRef]

applied sciences

MDPI

Article

Numerical Simulation of an InP Photonic Integrated Cross-Connect for Deep Neural Networks on Chip

Bin Shi *, Nicola Calabretta and Ripalta Stabile

Institute for Photonic Integration, Eindhoven University of Technology, 5600 MB Eindhoven, The Netherlands; n.calabretta@tue.nl (N.C.); r.stabile@tue.nl (R.S.)

* Correspondence: b.shi1@tue.nl

Received: 30 November 2019; Accepted: 26 December 2019; Published: 9 January 2020

Abstract: We propose a novel photonic accelerator architecture based on a broadcast-and-weight approach for a deep neural network through a photonic integrated cross-connect. The single neuron and the complete neural network operation are numerically simulated. The weight calibration and weighted addition are reproduced and demonstrated to behave as in the experimental measurements. A dynamic range higher than 25 dB is predicted, in line with the measurements. The weighted addition operation is also simulated and analyzed as a function of the optical crosstalk and the number of input colors involved. In particular, while an increase in optical crosstalk negatively influences the simulated error, a greater number of channels results in better performance. The iris flower classification problem is solved by implementing the weight matrix of a trained three-layer deep neural network. The performance of the corresponding photonic implementation is numerically investigated by tuning the optical crosstalk and waveguide loss, in order to anticipate energy consumption per operation. The analysis of the prediction error as a function of the optical crosstalk per layer suggests that the first layer is essential to the final accuracy. The ultimate accuracy shows a quasi-linear dependence between the prediction accuracy and the errors per layer for a normalized root mean square error lower than 0.09, suggesting that there is a maximum level of error permitted at the first layer for guaranteeing a final accuracy higher than 89%. However, it is still possible to find good local minima even for an error higher than 0.09, due to the stochastic nature of the network we are analyzing. Lower levels of path losses allow for half the power consumption at the matrix multiplication unit, for the same error level, offering opportunities for further improved performance. The good agreement between the simulations and the experiments offers a solid base for studying the scalability of this kind of network.

Keywords: artificial neural networks; deep neural network; image classification; photonic integrated circuits; semiconductor optical amplifiers; photonic neural network

1. Introduction

The boost in data volume of the information transient and data storage continuously stimulates the demand for high-speed information processing [1,2]. Artificial neural networks (ANNs) are becoming essential for feature extraction [3], image classification [4], time series prediction [5] and system optimization [6] as they are able to extract meaningful information from huge datasets more efficiently. They are also widely adopted by scientific communities to investigate bio-structure prediction [7], astronomical pattern extraction [8], nuclear fusion environment control [9], in telecommunication [10], etc. Novel neural network architectures based on non-von Neuman architectures to perform parallel computation have been demonstrated based on advanced electronics. As some examples, IBM TrueNorth [11], Neurogrid [12], SpiNNaker [13], and BrainDrop [14] are designed for spiking neural networks, while FPGA [15], EIE [16] and Google TPU [17] are for deep neural networks. The level of energy efficiency has been reported to be in the order of a few pJ/operation. However, the computation speed is constrained by the limited bandwidth of the electrical interconnections. Photonics technology provides a promising approach for neural network implementation as it offers parallel information processing when exploiting different domains (wavelength, polarization, phase, space), resulting in ultrabroad bandwidth that outperforms the electronics, while it decouples power consumption from computational speed. Recently, an ultrafast leaky integrate-and-fire neuron with a fiber-based system has been employed for spiking processing [18]. Large-scale optical neural networks using discrete optical components and micro-optics [19] and delay-based recurrent neural networks exploiting laser dynamics [20] have been reported. However, path-dependent and phase difference make the bulky systems difficult to scale up. Today's photonic integration technology can provide mature miniaturized solutions for high-performance sophisticated integrated circuits [21,22]. A photonic reservoir computing unit has been proposed based on time delays and semiconductor optical amplifiers (SOAs) [23] or Mach–Zehnder interferometers (MZIs) [24] for time-sequential recognition, though they are not programmable as they rely on distributed nonlinearities in the system. A photonic programable feed-forward neural network has been proposed based on a coherent approach using MZI elements [25], in which the optical neuron layer combines serval serial stages, resulting in phase noise accumulation. Micro-ring resonator-based optical neural networks with wavelength division multiplexing (WDM) operation have promised to increase interconnection bandwidth [26], however thermal crosstalk and low dynamic range complicate the weight calibration. Recently we have demonstrated the implementation of a photonic deep neural network (PDNN) via cross-connect circuits based on a broadcast-and-weight architecture, using SOAs and array waveguide gratings (AWGs) [27]. By running an image classification problem, we have demonstrated that an accuracy of up to 85.8% is possible. But, the influence of chip losses and optical crosstalk on the ultimate prediction accuracy has not been investigated yet. This is an important step to make for further improvement and scalability investigation.

In this work, we introduce the cross-connect-based photonic deep neural network and we simulate the matrix multiplication unit (MMU) via the *VPIphotonics Design Suit* (VPIphotonics, Berlin, Germany) simulation software. In particular, we benchmark the simulation results versus the experimental results to offer a solid platform for further analysis. We study the influence of the optical crosstalk, coming from the AWGs, as well as the impact of the path loss, to identify margins for further scalability per layer and energy saving. The single neuron and complete neural network operation are numerically simulated to provide guidelines on how to design future cross-connect photonic integrated chips for accelerating computation on-chip. In Section 2, we introduce the exploited SOA-based PDNN. The implementation and simulation with an optical cross-connect structure are described in Section 3, while the weighted calibration and neuron weighted addition are demonstrated in Section 4. The three-layer PDNN is used to solve the image classification problem in Section 5, followed by the conclusions in Section 6.

2. Photonic Deep Neural Network with Weight-SOAs

The implementation of deep neural networks via a photonic approach takes advantage of the available parallelism of light beams. Figure 1 depicts the envisaged photonic deep neural network which uses wavelength division multiplexing (WDM) input signals, from the photonic neuron to the large-scale neural network. Here in particular, we realize multiple weighted additions, linear operations in an artificial neuron, via a broadcast-and-weight architecture, which are the most computational heavy elements in the neural network.

The basic element of the neural network is an artificial neuron. Figure 1a depicts the basic neuron model with the output signal being $y_j = f(\sum W_{ij}\, x_i + b_j)$, where f is the activation function, x_i is the ith element of the input vector, W_{ij} is the weight factor for the input value x_i and b_j is the bias in the jth neuron, with the weighted addition given by $\sum W_{ij} x_i$. The output of one full layer of M neurons can be expressed as a vector: $y = f(W{\cdot}x + b)$, where x is the input vector with N elements,

W is the M × N weight matrix, b, a bias vector with M elements, with matrix multiplication $W \cdot x$. Figure 1b illustrates the corresponding photonic implementation with SOAs. In this instance, the input x is encoded onto several channels at different wavelengths and the individual input is weighted with the given gain/attenuation provided by an SOA. The weighted signals are then combined into a WDM signal and sent to the nonlinear function to provide a single wavelength neuron output. The nonlinear activation function can be realized in several ways, e.g., by employing the combination of a photodetector and a modulator [26], saturable absorbers [25], excitable lasers [28,29], wavelength converters [30], and phase change materials [31]. In this simulation work, we use a photodetector and off-line processing for nonlinear function and we mainly focus on the operation of weighted addition for the matrix multiplication. Utilizing photodetectors at the output of the matrix multiplication, the detected summation of all the weighted signals results in

$$V = (R \cdot Z_0/v_\pi) \cdot x \cdot \exp\,[h \cdot (I)]$$

where R is the signal detection response, assumed to be constant for dense WDM signals, Z_0 is the PD characteristic impedance and v_π the voltage at π phase shift. The vector $h\,(I)$ has N elements, the ith element $h(I_i)$ is the gain integrated over the length of the SOA for weighting input x_i, where the injection current is I_i. The outputs are then sent to the nonlinear function which processes the signal and produces the outputs of the neuron.

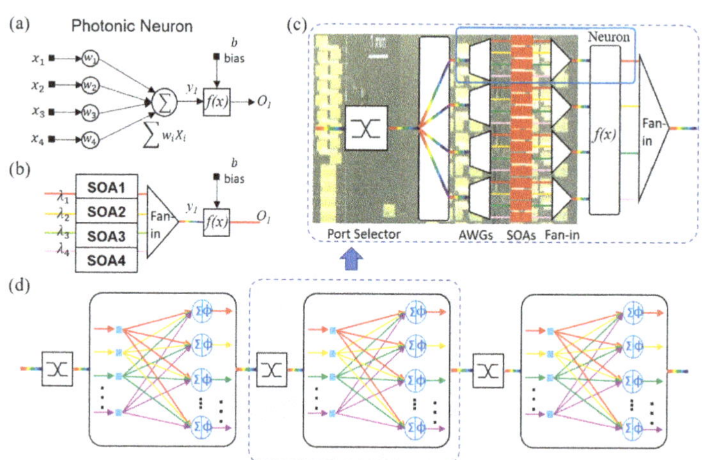

Figure 1. Photonic deep neural network based on the broadcast-and-weight architecture. (**a**) The artificial neuron model. (**b**) The implementation via arrays of semiconductor optical amplifiers (SOAs). (**c**) One full layer of neurons by exploiting one wavelength division multiplexing (WDM) input, with a shaded photonic integrated circuit micrography at the back, to underline that part of the circuitry that is realized on chip. (**d**) Scheme of a three-layer photonic deep neural network. The included port selector may be used to select the desired input source.

One full neural layer consists of linear matrix multiplications and nonlinear operations. The details of a neuron layer with four neurons, as used in this paper, are illustrated in Figure 1c, where the input WDM signal is selected by using a port selector that directs the desired input signal to this layer to be processed (see chip picture). The input signal is split and sent to the neurons (one neuron highlighted with a blue box). The AWG in the neuron de-multiplexes the input into individual channels, whose weight is assigned with different gains by using different weighted SOAs as shown in Figure 1b. The combined weighted signals from the four output ports pass through the activation function f, which is implemented via software with a hyperbolic tangent function. The output of the nonlinear activation function is a monochromatic wave that carries the information after the nonlinear operation. The outputs from different neurons in this layer are combined to be sent to the next layer of neurons for deeper processing. Figure 1d shows a schematic of the implementation of a full three-layer photonic deep neural network. The input of the neuron layers comes from the combined WDM output from the previous layer. The gray box shows one of the layers of the PDNN. By feeding forward the processed signals, the photonic deep neural network layer is realized. The included port selector may be used to select the desired input source.

To verify this photonic neural network concept, the simulation of the weight tuning and four channel weighted addition of a single photonic neuron is carried out and compared with the experimental results for calibration. The complete three-layer network is then implemented for solving the iris flower classification problem. A detailed analysis of the influence of the optical crosstalk and path losses on the error at each layer and on the final prediction accuracy is also executed to understand opportunities for improvements and scalability.

3. Optical Cross-Connect: Implementation and Simulation

We use *VPIphotonics* to simulate the integrated cross-connect-based weighted addition as the basic function of the photonic deep neural network. This software allows for numerical modeling of photonic systems as well as of photonic components within the integrated chips and for different material platforms. The simulated set up is built with symbolic blocks and a hierarchal structure. For the passive elements, we execute the simulation in frequency domain, while for the active elements, such as the SOAs, the transmission-line model is applied to model them in time domain [32].

The implemented and simulated setup scheme is showed in Figure 2. Figure 2a is the complete setup scheme for examining our cross-connect photonic integrated chip shown in Figure 1c, with similar operating conditions as in the real experiment, for analyzing the integrated SOA-based PDNN. The photonic integrated chip is an $8 \times 8 \times 8\lambda$ cross-connect, but in the experiments, a WDM input is used which contains 4 channels. An arbitrary waveform generator (detailed scheme shown in Figure 2b) is utilized to generate the electrical signal from the data file at 10 GSymbol/s, with 4 DACs with 8-bit precision. Figure 2c shows 4 lasers and 4 modulators for the optical signal generation of 4 input channels. The WDM input of four channels is generated via these four Mach–Zehnder interferometer-based modulators, with the electrical RF signal coming from the arbitrary waveform generator, and CW lasers at 193.1 THz, 193.5 THz, 193.9 THz, and 194.3 THz. A channel separation of 3.2 nm is used to match the channel separation of the AWG on chip. The input signal is coupled into the photonic matrix multiplication unit (MMU) with a 0 dBm optical input peak power for each channel. The output of the MMU is coupled to the receiver, shown in Figure 2d, which consists of a pre-amplifier with a noise figure of 5.0 dB, an AC-coupled (i.e., with DC-removing block in the simulation) 10 GHz avalanche photodetector (APD), and an analog-digital converter (ADC). The output from the MMU is then coupled to a 0.08 nm optical passband filter to monitor the peak power of one single channel at the output. The details of the schematic of part of the photonic MMU, i.e., the weighted addition unit, are illustrated in Figure 2e for the weighted addition demonstration. This will be used as the weighted addition part within a three-layer PDNN for demonstrating the iris flower classification. The path loss is the attenuation of the optical signal happening along the waveguide. The input signal is amplified with a pre-SOA and is split into 8 as for 8 neurons. Firstly, we study the performance of one neuron so that only one path carrying one WDM input signal is connected to the next SOA, the input vector selection SOA, that acts as a port selector as shown in Figure 1b. The WDM signal is then demultiplexed by an AWG, and the individual channel is weighted by the weight-SOA, and combined at the output of the unit. The parameters used in the simulation for the SOAs are listed in Table 1. The results are reported and explained, as related to the weight calibration and the weighted addition (Section 4), and the Iris classification application (Section 5) together with the analysis of the impact of the optical crosstalk and the optical path loss.

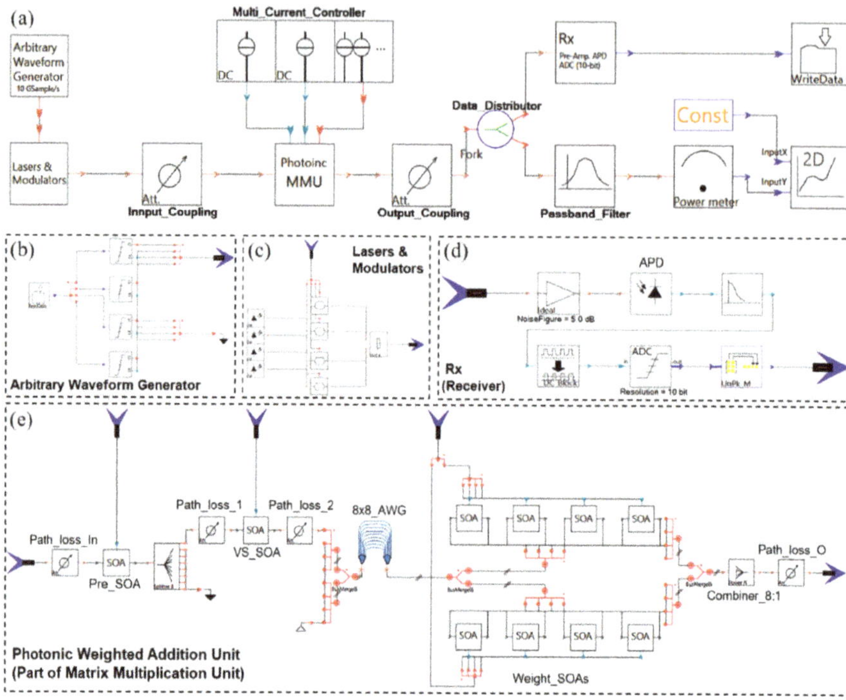

Figure 2. Photonic deep neural network (PDNN) simulation scheme on software *VPIphotonics*. (**a**) System for examining the PDNN. (**b**) Arbitrary waveform generator. (**c**) Lasers and modulators. (**d**) Receiver. (**e**) One photonic weighted addition unit (part of the matrix multiplication unit, MMU).

Table 1. The parameters used in the simulation of SOA.

Parameters	Value	Unit	Parameters	Value	Unit
Device Section Length	1000×10^{-6}	m	Nonlinear Gain Coefficient	1.0×10^{-23}	m^3
Active Region Type	MQW		Nonlinear Gain Time Constant	5.00×10^{-13}	s
Active Region Width	2.0×10^{-6}	m	Carrier Density Transparency	1.0×10^{24}	m^{-3}
Active Region Thickness	250×10^{-6}	m	Linear Recombination	1.0×10^{8}	s^{-1}
Active Region Thickness MQW	100×10^{-6}	m	Bimolecular Recombination	1.0×10^{-16}	m^3/s
Active Region Thickness SCH	200×10^{-6}	m	Auger Recombination	2.1×10^{-41}	m^6/s
Current Injection Efficiency	1		Carrier Capture Time Constant	3.0×10^{-11}	s
Nominal Frequency	193.7	THz	Carrier Escape Time Constant	1.0×10^{-10}	s
Group Index	3.52		Initial Carrier Density	8.0×10^{23}	m^{-3}
Polarization Model	TE		Chirp Model	Linewidth Factor	
Internal Loss	3000	m^{-1}	Linewidth Factor	3	
Confinement Factor	0.3		Linewidth Factor MQW	3	
Confinement Factor MQW	0.07		Differential Index	-1.0×10^{-26}	m^3
Confinement Factor SCH	0.56		Differential Index MQW	-1.0×10^{-26}	m^3
Gain Shape Model	Flat		Differential Index SCH	-1.5×10^{-26}	m^3
Gain Model	Logarithmic		Carrier Density Ref. Index	1.0×10^{24}	m^{-3}
Gain Coefficient Linear	4.00×10^{-20}	m^2	Noise Model	Inversion Parameter	
Gain Coefficient Logarithmic	6.9×10^{4}	m^{-1}	Inversion Parameter	1.2	

4. Implementation of Weight Calibration and Weighted Addition

For the operation of the SOA-based photonic neural network, a calibration of the weighting is required for correctly assigning the given weight factors to the input data. For this simulation implementation, the weight-SOAs are identical for all the input channels so that we demonstrate the weight calibration on one of the input channels. For the weight calibration, the input can be a non-return-to-zero on-off keying (NRZ OOK) signal or multi-level data input. As the weighting of the input data is performed after the AWG, the fixed optical crosstalk from the AWG will influence the output optical signal. We consider two extreme conditions for the optical crosstalk level: when switching ON (injection current at 70 mA) all the weight-SOAs, the optical crosstalk coming from the adjacent channels is expected to be maximum ($XTalk_{max}$), while when all the weight-SOAs are OFF (zero injected current), the induced optical crosstalk by that the corresponding channels will be the minimum ($XTalk_{min}$). Due to the complexity of operation conditions, we consider the average between these two scenarios in order to generate the weight control curve, in order to minimize the error induced by the optical interference.

The crosstalk in the AWG in the photonic MMU (see Figure 2e) is set at −20 dB, as experimentally measured in [33]. Firstly, when all the weight-SOAs are set to OFF ($XTalk_{min}$), but one of these weight-SOAs is injected with currents from 0 mA to 70 mA, we record the signal peak power at channel 1, 193.1 THz, from the monitoring power meter as shown in Figure 2a. Then, we also record the signal power when all the weight-SOAs are set to ON ($XTalk_{max}$), and one of these weight-SOAs is injected with currents from 0 mA to 70 mA. The blue and red solid lines in Figure 3a plot the simulated result in the condition of $XTalk_{min}$ and $XTalk_{max}$, respectively. For comparison, we also superimpose the measured curves in both cases: the blue crosses curve represents the experiment points with all SOAs ON, and the red triangles curve plots the experimental results with all SOAs OFF.

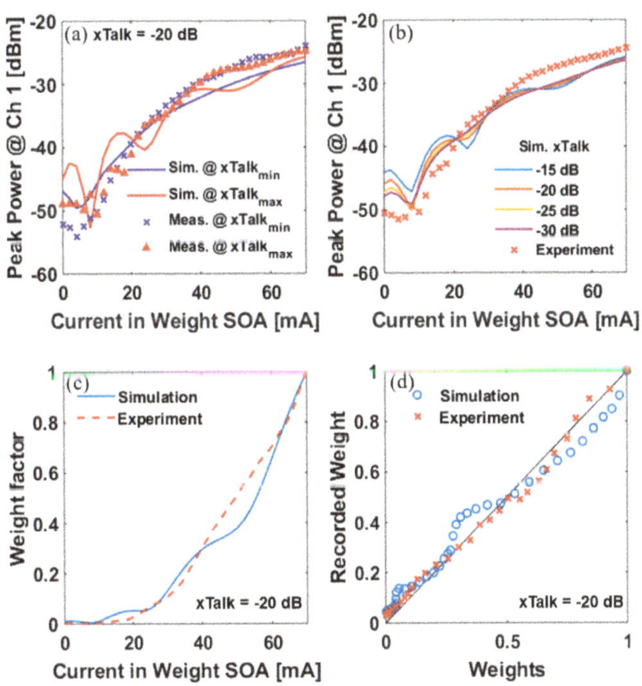

Figure 3. Weight calibration. (**a**) Peak power of channel at 193.1 THz with minimum crosstalk (blue) and maximum crosstalk (red), in simulation (solid line), or experiment (cross/triangle points), versus the injection current at the weight-SOA. (**b**) Mean peak power of channel at 193.1 THz from the two curves obtained in (**a**), and with simulated crosstalk at −15 dB, −20 dB, −25 dB and −30 dB. (**c**) Weight control curves, in simulation (solid line) and experiment (dash line), with crosstalk of −20 dB and reference power level at −25 dBm. (**d**) Correlation between the weight assigned by the weight-SOA and the obtained weight at the output, in simulation (blue circles) and experiments (red crosses). The black line is a reference line for perfect matching.

The curve trends, in the case of simulation and experimental results, are very similar. We then scan the optical crosstalk level to investigate the influence of the optical crosstalk on the peak power curves. In Figure 3b, the blue, red, yellow and violet solid lines show the peak power on channel 1 (averaged from the curves shown in Figure 3a), where the simulated crosstalk for the AWG is set to −15 dB, −20 dB, −25 dB and −30 dB, respectively. It is visible that higher crosstalk will induce greater oscillation when tuning the injection current at the weight-SOA. The oscillation might be due to the interference between the crosstalk and the signal in the desired path. The experimental result is also presented with red crosses, which is the mean value of the experimental results shown in Figure 3a. The plots indicate that a dynamic range wider than 25 dB is possible. The slight difference between simulations and experiments may be attributed to the difference in gain efficiency as hypothesized for the SOA modeling. The weight control curve in Figure 3c is generated by the power control curves in Figure 3b, with reference weight '1' level at −25 dBm optical input power, which is the signal peak power when injection current of the weight-SOA is set at 70 mA. The weight calibration curves show two semi-linear operation regimes, both for the simulation (Figure 3c, blue solid line) and the experiment (Figure 3c, red dashed line). These two regions correspond to the two different SOA operation regimes: the transparency operation and linear amplification. After the weight calibration, we obtain the correlation between the assigned weight and the obtained one for the simulated and the experimental operation in Figure 3d. An error lower than 0.12 for the simulation results is obtained, when compared to the reference perfect linear relation as shown by the black line.

The weighted addition corresponds to the linear operation part in a neuron. The performance of the weighted addition is of importance for the signal processing in a neural network. To estimate the impairments induced by the weighted addition, we calculate the normalized root mean square error (NRMSE), i.e., the discrepancy between the measured data and the expected data. We use the calibrated weight control curve to set the weight factors for different input channels, and calculate the NRMSE while tuning the weight factor from 0 to 1. Figure 4a plots the results of two-channel weighted addition, where channel 2 is fixed to the weight '1', while the weight for channel 1 is tuned over the overall range from 0 to 1. We also change the optical crosstalk in the chip to see the impact of the optical crosstalk on the weighted addition. The blue, red, yellow and violet lines show the error changes when the crosstalk is set at −15 dB, −20 dB, −25 dB and −30 dB, respectively. The shaded area shows the error range obtained from the experiments for two-channel addition. The error variation is attributed to the calibration of the weight control, as already anticipated in the weight factor curve in Figure 3c. The error related to the weighted addition operation increases when the induced optical crosstalk is greater, as a high level of crosstalk eventually results in a lower dynamic range. The same high crosstalk level enhances the peak power oscillation recorded for generating the weight control curve, resulting in severe error variations, as already anticipated in Figure 3d. Nevertheless, this fits perfectly within the error variation window we found for our experimental results (see the dashed box in Figure 4a). The same analysis is done while changing the number of channels added to the WDM input. Figure 4b,c plot the resulting errors for three- and four-channel weighted additions, respectively. A visibly smaller error is presented for three-channel and four-channel weighted additions. The effect of the optical crosstalk is reduced as the oscillation power caused by the optical crosstalk is relatively smaller with respect to the dominating signal power coming from the addition of all the input multiple signals. This suggests that the higher the number of inputs into the neuron, the better the accuracy when operating within the available power budget. Finally, Figure 4d summarizes the obtained results, by plotting the maximum errors versus the number of channels in weighted addition.

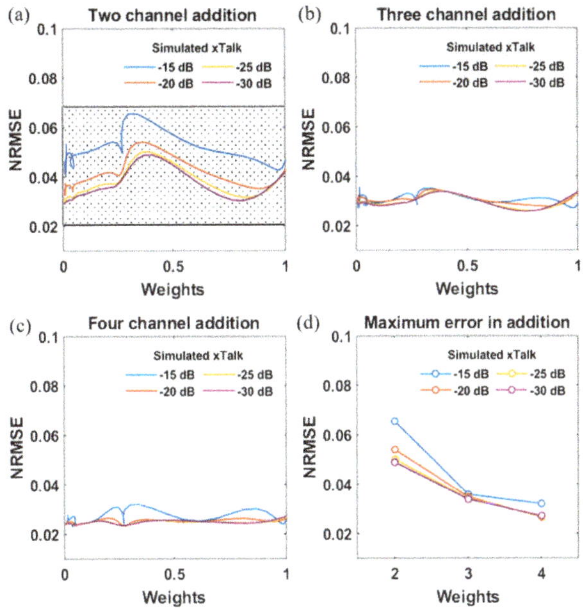

Figure 4. (**a**) Two-channel weighted addition, (**b**) three-channel addition, and (**c**) four-channel addition when tuning weights in channel 1 and fixed weight on other channels; (**d**) Maximum error versus the number of channels in weighted addition. Optical crosstalk levels are −15 dB (blue), −20 dB (red), −25 dB (yellow) and −30 dB (violet).

5. Image Classification via a Three-Layer Photonic Deep Neural Network

To investigate the performance of a complete neural network based on the combination of the AWG and SOA technology, for a broadcast-and-weight architecture, we implement and simulate an image classification problem, namely the iris flower classification problem, which has been reported to be able to be solved by using a deep neural network (DNN). The iris database includes three classes (Setosa, Versicolor, and Virginica) of 50 instances each [34]. Per each instance, the iris flower category is identified by observing four of its attributes: length and width of its sepals and petals. For this demonstration, we have executed the training of this DNN via the simulation platform *Tensorflow* [35], where we have used 120 instances as a training database. In order to make use of 4 weighted addition circuitries already available on chip and per layer, a feed-forward network made of 2 hidden layers with 4 neurons each and an output layer with 3 neurons (see Figure 5b), is trained on a computer. The attributes are encoded into 2^6 optical power levels at the photonic MMU input. The trained weight matrix is mapped to the matrix multiplication on the photonic components. The simulated structure of one layer of neurons is shown in Figure 5a, which is used to replace the photonic matrix multiplication unit in Figure 2e. The same chip is indeed capable of eight channel inputs, but we used four inputs for this classification problem. A total of 16 weight-SOAs in this matrix multiplication unit are used to assign the trained weight matrix from the trained DNN model to the PDNN. The hyperbolic tangent activation function is implemented offline after the O/E conversion. The output from the first hidden layer serves as input to the second layer (via the arbitrary waveform generator) and the output from the second hidden layer serves as input to the third (output) layer. Finally, the output of the third layer, after the *SoftMax* transfer function, $P(y=j) = e^{y_j} / \sum_{j=1}^{n} e^{y_j}$, provides the predicted probability of the output samples y of belonging to class j. Figure 5c presents the output data at the output (i) of the 1st neuron in the 1st layer, (ii) of the 2nd layer and (iii) of the 3rd layer, with the blue line being the simulation results and red line the expectations and resulting in errors of 0.123, 0.051 and 0.055 respectively. These errors represent the performance of layers of photonic neurons. Higher error at the first layer may be due to the high optical signal noise ratio (OSNR) required multilevel encoding of the input signal, while a better performance at the 2nd and 3rd layer is attributed to the filtering of the signal level into lower levels after the first hidden layer. Also, the output of the first layer appears to be the most important for this classification problem, and therefore the utilization of three layers is slightly overstated.

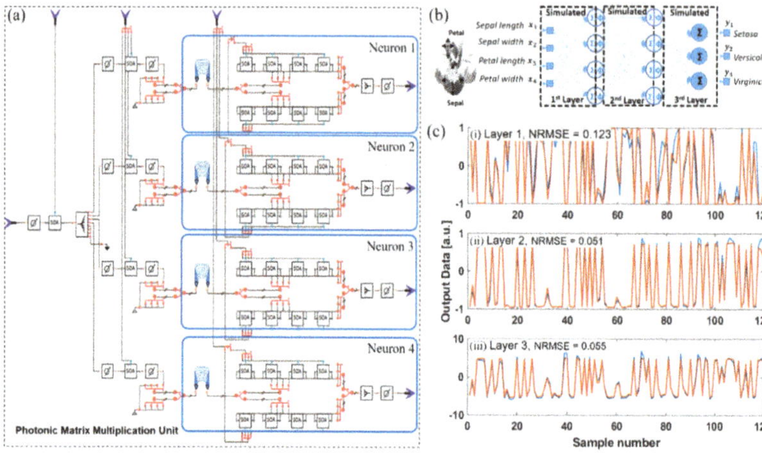

Figure 5. (**a**) The simulation structure of one layer of neurons; (**b**) Trained 3-layer deep neural network (DNN) employed to solve the iris flower classification. (**c**) Output data obtained from Neuron 1 at (i) Layer 1, (ii) Layer 2, and (iii) Layer 3, with calculated errors between simulated computation (blue line) and the expected computation (red line), resulting in errors of 0.123, 0.051, and 0.055, respectively.

The correlation matrix between the prediction and the labels of the samples is used to show the final accuracy obtained via the multilayer photonic neural network (see Figure 6). We consider three cases for the sake of understanding the influence of the photonic layer implementation. In Figure 6a we display the prediction accuracy as coming from the trained DNN on a PC. This is calculated to be 95% since 6 out of 120 iris flower instances are falsely predicted. We simulate the DNN after adding, time by time, the photonic deep neuron network layers. The prediction accuracy decreases as the number of layers of the photonic neural network increases. The accuracy changes from 89.7% when the 1st layer is substituted with a photonic layer (Figure 6b), down to 86.7% when both the 1st and the 2nd layers are substituted with photonic layers (Figure 6c), and down to 85.8% when all 3 layers of matrix multiplications are computed via three photonic layers (Figure 6d). This may be due to the error accumulation which causes prediction accuracy degradation. Furthermore, the simulation result aligns well with the experimental result trend as shown in Figure 7a.

Figure 7a plots the error evolution with an increasing number of the photonic layers. The solid lines with open symbols represent the results from the simulation and the dashed lines with filled symbols represent the experimental results. The circles show the error induced by each single layer on the 3-layer network, where the errors keep staying almost at the same level, about 0.07 in the simulation and 0.08 in the experiment. The triangles plot the accumulated error from layer to layer, which increases from 0.1 to 0.18 for simulation and from 0.10 to 0.20 in the experiment. The squares represent the final prediction accuracy as we calculated from the correlation matrix, which decreases from 89.2% to 85.8% as shown in Figure 6 for simulation and from 91.2% to 85.8% for experiments [27]. The experimental results show great agreement with the simulations, which means that investigating the performance while changing some of the parameters involved in the photonic integrated circuit (PIC) will help to get some insight into the photonic chip architecture and scalability. From the perspective of the final prediction accuracy and error induced by the photonic neural network chip, the impact of the optical crosstalk from the AWG and the waveguide crossings are investigated. We tune the crosstalk from −15 dB to −30 dB with 1 dB steps and implement the 3-layer neural network after generating the weight calibration curves as reported in Section 4. Figure 7b plots the results at the output of layer 1, with the blue line representing the average NRMSE from 4 neurons at layer 1 and the red line plotting the variation of the final prediction. Similarly, Figure 7c,d illustrate the average NRMSE and the final prediction versus the optical crosstalk in the layers. The error induced by the chip is almost in the same range for different crosstalk values, though it slightly reduces when the crosstalk decreases. The prediction accuracy for Layer 1 in Figure 7b shows a stronger crosstalk dependency; a smaller optical crosstalk at Layer 1 provides a better prediction accuracy. This may be related to the fact that the first layer operates on high resolution multilevel input signals, which require a higher optical to signal ratio available. A better accuracy also appears when the crosstalk is high, i.e., near −15 dB. This might be attributed to the errors leading the prediction of the flower label to a different minima location, i.e., to changes of the state of the network, as will also be found in Section 5. Figure 7c shows a flattened accuracy for optical crosstalk smaller than −20 dB. Figure 7d shows an even more flattened accuracy level as the variation of the induced error is smaller. The accuracy level is maintained from the 2nd layer onwards.

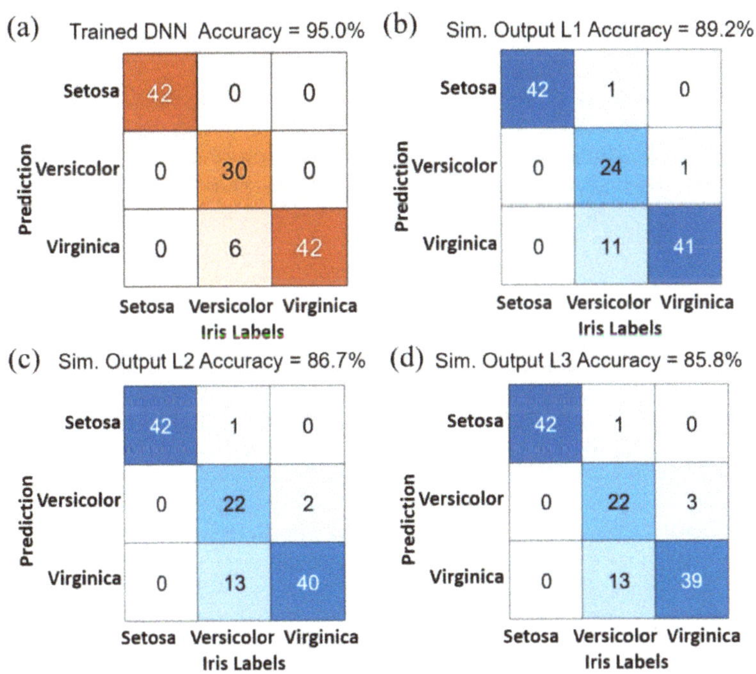

Figure 6. (**a**) Label prediction of the trained DNN, indicating an accuracy of 95%. (**b**) Simulated image prediction using photonic DNN as the 1st hidden layer, with an accuracy of 89.2%. (**c**) Simulated label prediction using photonic DNN as the 1st and 2nd hidden layers, with an accuracy of 86.7%. (**d**) Simulated label prediction of the 3-layer photonic DNN, with an accuracy of 85.8%.

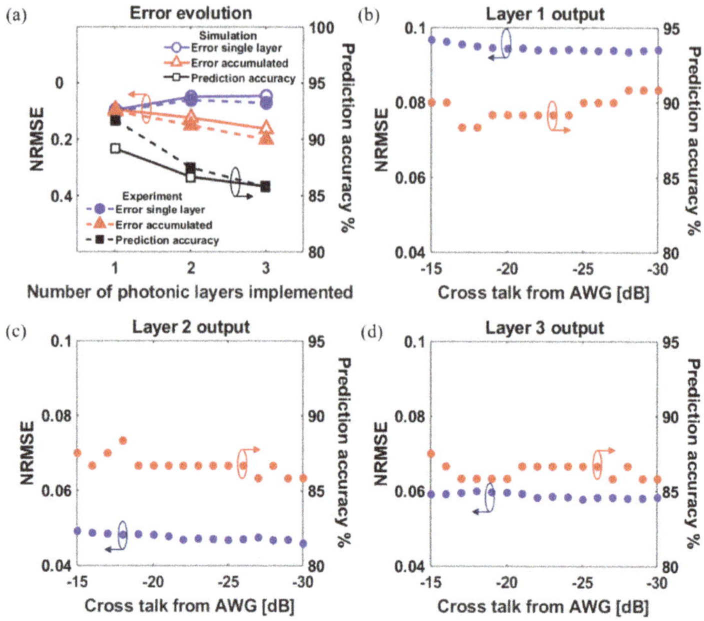

Figure 7. Error evolution: (**a**) Normalized root mean square error (NRMSE) versus the number of implemented photonic layers in simulation (solid line filled points) and experiment (dashed line open points), from single photonic layer (circles), the accumulation (triangles) and the corresponding prediction accuracy (squares). Crosstalk tuning: The induced error (blue circles) and the final prediction accuracy (red circles) versus the crosstalk from AWGs, recorded simulation results from (**b**) output of layer 1, (**c**) output of layer 2, and (**d**) output of layer 3.

5.1. Energy Consumption Versus Physical Layer Impairments

The performance of the PDNN is expected to be influenced also by the reference energy level used to operate the optical engine and by the loss on chip. Therefore, we study the performances of the PDNN by executing the iris classification problem, while tuning the reference power level of weight factor '1' used in the integrated circuit, i.e., while tuning the current used in the SOAs, as well as while assuming different waveguide losses for the optical paths. This analysis is carried out to understand opportunities for energy savings and best chip physical layer characteristics, which still guarantee a high level of prediction accuracy. In particular, for this analysis, we consider only the waveguide loss as the main loss component as this is true for large size PIC. Therefore, we calculate the NRMSE at the output of each photonic neuron layer, as well as the prediction accuracy obtained when involving this layer in the 3-layer DNN for the iris flower classification, and we provide 3D color maps of error and accuracy as a function of the scanning waveguide losses and energy consumption on different reference power levels. Figure 8a illustrates the average errors obtained at the output of the 1st layer, the 2nd layer and the 3rd layer as a function of the waveguide loss and the energy consumption. It can be observed that fewer losses allow less energy consumption, for the same error level. This suggests that by only improving the waveguide loss on chip we can double the energy savings. The induced error from the photonic DNN is expected to be greater when the waveguide losses are higher and the energy consumption per operation at the matrix multiplication unit is lower, as the dynamic range is not enough to be able to distinguish multilevel data. On the contrary, smaller error values are observed with lower waveguide loss and higher energy used for the weighted addition operations (see moving from lighter color to darker color, from bottom right to top left side in Figure 8a and for each layer). For Layer 1 this is more evident and is coherent with our previous conclusions. It is not surprising that we need to either tune the reference power to higher levels or reduce the waveguide loss to obtain smaller errors at the neuron signal processing.

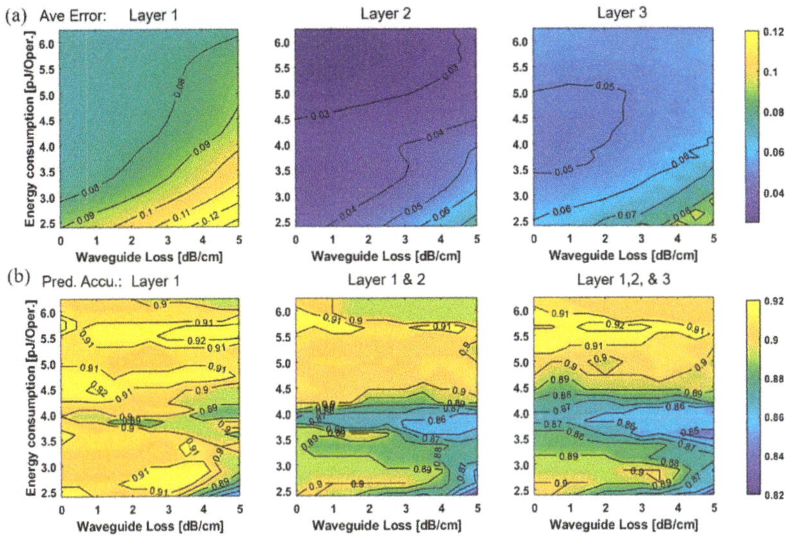

Figure 8. Investigation performance of the PDNN on computing energy and waveguide loss. (**a**) Calculated average NRMSE from output data obtained from Layers 1–3; (**b**) Corresponding prediction accuracy when Layer 1, Layers 1,2, and Layers 1–3 are implemented with photonic neuron layers.

Furthermore, the final prediction accuracies for the cases when Layer 1, Layers 1 and 2, and all three layers are implemented by using the photonic integrated chip are shown in Figure 8b. The yellow color corresponds to a higher prediction accuracy, while the blue color corresponds to a lower prediction accuracy. The prediction accuracy results do not show the same trend as shown in Figure 8a, i.e., the trend for the error induced on the layer operations, which indicates that an induced error is not necessarily reducing the performance of the photonic neural network. The result for Layer 1 shows that a good prediction is obtained for an error smaller than 0.09 and in that region the accuracy remains generally very stable, while the accuracy for higher error levels is variable, and generally worse. This suggests that there is a certain maximum level of error we should never cross at the first layer for always guaranteeing a good accuracy. The prediction mapping from the implementation of Layers 1 and 2 shows a slight decrease in the accuracy as the error accumulated from the previous layer. In the case of adding the contribution of Layer 3, it is the result of the small error accumulation as well, for the final prediction accuracy on this 3-layer photonic neural network system. However, it is evident that the two different regions are delineated when more error is accumulated from layer to layer. In particular, for the complete 3-layer photonic neural network, the best performance condition (accuracy = 92%) is found when the energy efficiency is around 5.6 pJ/operation and the waveguide loss ranges between 1.5 and 3.5 dB/cm. However it is possible to distinguish two areas where the accuracy is already higher than 89%: (1) the total energy consumption is above 4.5 pJ/operation, irrespective of the path loss; (2) the area at the left down corner where the averaged energy consumption is around 2.8 pJ/operation and the loss covers almost the full considered range (up to 4 dB/cm). The region (1) performs well due to a higher signal power with higher power consumption on the system with smaller errors induced and accumulated. We believe the region (2) appears due to the presence of more local minima, whose presence is determined by the combination of path losses, power level and optical crosstalk. Furthermore, the level of noise present in the network makes it a stochastic network where the intrinsic noise is supposed to provide better accuracy. Noise might play a positive role for low power levels as a good prediction is presented. However, this behavior has to be further explored for quantification. The identification of small error regions and their slight influence on the final prediction accuracy, as well as the maximum level of error at the first layer shown in Figure 8 suggests that the PDNN might be further scaled up, with prior physical parameters and error optimization.

6. Conclusions

We propose a photonic deep neural network based on the use of WDM input signals, and an SOA-based matrix multiplication unit. The integrated photonic neural network is employed as a weighted addition, thereafter combined with an offline hyperbolic tangent function as a nonlinear function is demonstrated on the simulation platform *VPIphotonics*. We study the weight calibration and weighted addition with different crosstalk of photonic integrated AWGs and SOA-based cross-connects. The error from the weighted addition is found to decrease when the numbers of input channels increase, so that a high number of input channels is beneficial for the implementation of the PDNN. A trained 3-layer DNN is implemented by reconfiguring the weight setting on the subnetwork and feeding the layer output to the next layer. The performance is simulated with different values of crosstalk, energy consumption per operation, and waveguide loss. The experimental results are in agreement with the simulation results, meaning that the implemented simulation offers a solid base for further study of scalability for this kind of network architecture. The results show that the photonic DNN is robust to the noise added during the signal processing. The error induced by the first layer is greater than the next two layers, due to the higher resolution multilevel encoding at the input layer with respect to the resolution at the 2nd and 3rd layer, but the error is not necessarily degrading the performance for a maximum allowed error. The performance analysis as a function of the path losses suggests the photonic neural network could be further optimized for lower power consumption. These results provide enough insights for the design of scalable photonic neural networks to a higher dimension for solving higher complexity problems. Finally, in future, a combination of the weighted addition function with on-chip non-linearities holds the promise to enable further acceleration for computation.

Author Contributions: Conceptualization, B.S., N.C. and R.S.; formal analysis, B.S. and R.S.; investigation, B.S.; methodology, B.S., N.C. and R.S.; validation, B.S.; visualization, B.S.; writing—review & editing, B.S., N.C. and R.S. All authors have read and agreed to the published version of the manuscript.

Funding: This work is financially supported by the Netherlands Organization of Scientific Research (NWO) under the Gravitation program (Zwaartekracht programma), 'Research Centre for Integrated Nanophotonics'.

Acknowledgments: The authors thank the technical support from VPIphotonics.

Conflicts of Interest: The authors declare no conflict of interest.

References

1. McAfee, A.; Brynjolfsson, E. Big Data: the Management Revolution. *Harv. Bus. Rev.* **2012**, *90*, 60–66, 68, 128. [PubMed]
2. Philip Chen, C.L.; Zhang, C.-Y. Data-Intensive Applications, Challenges, Techniques and Technologies: A Survey on Big Data. *Inf. Sci. (Ny)* **2014**, *275*, 314–347. [CrossRef]
3. Masci, J.; Meier, U.; Cire, D. Stacked Convolutional Auto-Encoders for Hierarchical Feature Extraction. In *Proceedings of the Artificial Neural Networks and Machine Learning-ICANN 2011*; Honkela, T., Duch, W., Girolami, M., Kaski, S., Eds.; Springer: Berlin, Germany, 2011; pp. 52–59.
4. Lawrence, S.; Giles, C.L.; Tsoi, A.C.; Back, A.D. Face Recognition: A Convolutional Neural-Network Approach. *IEEE Trans. Neural Netw.* **1997**, *8*, 98–113. [CrossRef]
5. Hill, T.; O'Connor, M.; Remus, W. Neural Network Models for Time Series Forecasts. *Manag. Sci.* **1996**, *42*, 1082–1092. [CrossRef]
6. Chow, T.T.; Zhang, G.Q.; Lin, Z.; Song, C.L. Global Optimization of Absorption Chiller System by Genetic Algorithm and Neural Network. *Energy Build.* **2002**, *34*, 103–109. [CrossRef]

7. Zeng, H.; Edwards, M.D.; Liu, G.; Gifford, D.K. Convolutional Neural Network Architectures for Predicting DNA–Protein Binding. *Bioinformatics* **2016**, *32*, i121–i127. [CrossRef]
8. Ball, N.M.; Brunner, R.J. Data Mining and Machine Learning in Astronomy. *Int. J. Mod. Phys. D* **2010**, *19*, 1049–1106. [CrossRef]
9. Cannas, B.; Fanni, A.; Marongiu, E.; Sonato, P. Disruption Forecasting at JET Using Neural Networks. *Nucl. Fusion* **2004**, *44*, 68–76. [CrossRef]
10. Fischer, M.M.; Gopal, S. Artificial Neural Networks: A New Approach to Modeling Interregional Telecommunication Flows. *J. Reg. Sci.* **1994**, *34*, 503–527.
11. Akopyan, F.; Sawada, J.; Cassidy, A.; Alvarez-Icaza, R.; Arthur, J.; Merolla, P.; Imam, N.; Nakamura, Y.; Datta, P.; Nam, G.J.; et al. TrueNorth: Design and Tool Flow of a 65 mW 1 Million Neuron Programmable Neurosynaptic Chip. *IEEE Trans. Comput. Des. Integr. Circuits Syst.* **2015**, *34*, 1537–1557. [CrossRef]
12. Benjamin, B.V.; Gao, P.; McQuinn, E.; Choudhary, S.; Chandrasekaran, A.R.; Bussat, J.; Alvarez-Icaza, R.; Arthur, J.V.; Merolla, P.A.; Boahen, K. Neurogrid: A Mixed-Analog-Digital Multichip System for Large-Scale Neural Simulations. *Proc. IEEE* **2014**, *102*, 699–716. [CrossRef]
13. Furber, S.B.; Galluppi, F.; Temple, S.; Plana, L.A. The SpiNNaker Project. *Proc. IEEE* **2014**, *102*, 652–665. [CrossRef]
14. Neckar, A.; Fok, S.; Benjamin, B.V.; Stewart, T.C.; Oza, N.N.; Voelker, A.R.; Eliasmith, C.; Manohar, R.; Boahen, K. Braindrop: A Mixed-Signal Neuromorphic Architecture with a Dynamical Systems-Based Programming Model. *Proc. IEEE* **2019**, *107*, 144–164. [CrossRef]
15. Zhang, C.; Li, P.; Sun, G.; Guan, Y.; Xiao, B.; Cong, J. Optimizing FPGA-Based Accelerator Design for Deep Convolutional Neural Networks. In Proceedings of the FPGA 2015–2015 ACM/SIGDA International Symposium on Field-Programmable Gate Arrays, Monterey, CA, USA, 22–24 February 2015; ACM Press: New York, NY, USA, 2015; pp. 161–170.
16. Han, S.; Liu, X.; Mao, H.; Pu, J.; Pedram, A.; Horowitz, M.A.; Dally, W.J. EIE: Efficient Inference Engine on Compressed Deep Neural Network. In Proceedings of the 2016 ACM/IEEE 43rd Annual International Symposium on Computer Architecture (ISCA), Seoul, Korea, 18–22 June 2016; pp. 243–254.
17. Jouppi, N.P.; Borchers, A.; Boyle, R.; Cantin, P.; Chao, C.; Clark, C.; Coriell, J.; Daley, M.; Dau, M.; Dean, J.; et al. In-Datacenter Performance Analysis of a Tensor Processing Unit. In Proceedings of the Proceedings-International Symposium on Computer Architecture, Toronto, ON, Canada, 24–28 June 2017; ACM Press: New York, NY, USA, 2017; pp. 1–12.
18. Kravtsov, K.S.; Fok, M.P.; Prucnal, P.R.; Rosenbluth, D. Ultrafast All-Optical Implementation of a Leaky Integrate-and-Fire Neuron. *Opt. Express* **2011**, *19*, 2133–2147. [CrossRef]
19. Bueno, J.; Maktoobi, S.; Froehly, L.; Fischer, I.; Jacquot, M.; Larger, L.; Brunner, D. Reinforcement Learning in a Large-Scale Photonic Recurrent Neural Network. *Optica* **2018**, *5*, 756–760. [CrossRef]
20. Nakayama, J.; Kanno, K.; Uchida, A. Laser Dynamical Reservoir Computing with Consistency: An Approach of a Chaos Mask Signal. *Opt. Express* **2016**, *24*, 8679–8692. [CrossRef]
21. Stabile, R.; Rohit, A.; Williams, K.A. Monolithically Integrated 8 × 8 Space and Wavelength Selective Cross-Connect. *J. Light. Technol.* **2014**, *32*, 201–207. [CrossRef]
22. Smit, M.; Leijtens, X.; Ambrosius, H.; Bente, E.; van der Tol, J.; Smalbrugge, B.; de Vries, T.; Geluk, E.-J.; Bolk, J.; van Veldhoven, R.; et al. An Introduction to InP-Based Generic Integration Technology. *Semicond. Sci. Technol.* **2014**, *29*, 083001. [CrossRef]
23. Vandoorne, K.; Dambre, J.; Verstraeten, D.; Schrauwen, B.; Bienstman, P. Parallel Reservoir Computing Using Optical Amplifiers. *IEEE Trans. Neural Netw.* **2011**, *22*, 1469–1481. [CrossRef]
24. Vandoorne, K.; Mechet, P.; Van Vaerenbergh, T.; Fiers, M.; Morthier, G.; Verstraeten, D.; Schrauwen, B.; Dambre, J.; Bienstman, P. Experimental Demonstration of Reservoir Computing on a Silicon Photonics Chip. *Nat. Commun.* **2014**, *5*, 3541. [CrossRef]
25. Shen, Y.; Harris, N.C.; Skirlo, S.; Prabhu, M.; Baehr-Jones, T.; Hochberg, M.; Sun, X.; Zhao, S.; Larochelle, H.; Englund, D.; et al. Deep Learning with Coherent Nanophotonic Circuits. *Nat. Photonics* **2017**, *11*, 441–446. [CrossRef]
26. Tait, A.N.; De Lima, T.F.; Zhou, E.; Wu, A.X.; Nahmias, M.A.; Shastri, B.J.; Prucnal, P.R. Neuromorphic Photonic Networks Using Silicon Photonic Weight Banks. *Sci. Rep.* **2017**, *7*, 1–10. [CrossRef] [PubMed]
27. Shi, B.; Calabretta, N.; Stabile, R. Deep Neural Network Through an InP SOA-Based Photonic Integrated Cross-Connect. *IEEE J. Sel. Top. Quantum Electron.* **2020**, *26*, 1–11. [CrossRef]

28. Peng, H.-T.; Nahmias, M.A.; de Lima, T.F.; Tait, A.N.; Shastri, B.J. Neuromorphic Photonic Integrated Circuits. *IEEE J. Sel. Top. Quantum Electron.* **2018**, *24*, 1–15. [CrossRef]
29. Nahmias, M.A.; Peng, H.-T.; de Lima, T.F.; Huang, C.; Tait, A.N.; Shastri, B.J.; Prucnal, P.R. A TeraMAC Neuromorphic Photonic Processor. In Proceedings of the 2018 IEEE Photonics Conference (IPC), Reston, VA, USA, 30 September–4 October 2018; pp. 1–2.
30. Mourgias-Alexandris, G.; Tsakyridis, A.; Passalis, N.; Tefas, A.; Vyrsokinos, K.; Pleros, N. An all-Optical Neuron with Sigmoid Activation Function. *Opt. Express* **2019**, *27*, 9620. [CrossRef]
31. LeCun, Y.; Bottou, L.; Bengio, Y.; Haffner, P. Gradient-Based Learning Applied to Document Recognition. *Proc. IEEE* **1998**, *86*, 2278–2323. [CrossRef]
32. Lowery, A.J. Amplified Spontaneous Emission in Semiconductor Laser Amplifiers. Validity of the Transmission-Line Laser Model. *IEE Proc. Part J. Optoelectron.* **1990**, *137*, 241–247. [CrossRef]
33. Stabile, R.; Rohit, A.; Williams, K.A. Dynamic Multi-Path WDM Routing in a Monolithically Integrated 8×8 Cross-Connect. *Opt. Express* **2014**, *22*, 435–442. [CrossRef]
34. Fisher, R.A. The Use of Multipe Measurements in Taxonomic Problems. *Ann. Eugen.* **1936**, *7*, 179–188. [CrossRef]
35. Chan, T.-H.; Jia, K.; Gao, S.; Lu, J.; Zeng, Z.; Ma, Y. PCANet: A Simple Deep Learning Baseline for Image Classification? *IEEE Trans. Image Process.* **2015**, *24*, 5017–5032. [CrossRef]

applied sciences

MDPI

Article

LED Nonlinearity Estimation and Compensation in VLC Systems Using Probabilistic Bayesian Learning

Chen Chen [1,*], Xiong Deng [2], Yanbing Yang [3], Pengfei Du [4], Helin Yang [4] and Lifan Zhao [4]

1 School of Microelectronics and Communication Engineering, Chongqing University, Chongqing 400044, China
2 Department of Electrical Engineering, Eindhoven University of Technology (TU/e), Flux Building, 5600MB Eindhoven, The Netherlands
3 College of Computer Science, Sichuan University, Chengdu 610065, China
4 School of Electrical and Electronic Engineering, Nanyang Technological University, Singapore 639798, Singapore
* Correspondence: c.chen@cqu.edu.cn

Received: 27 May 2019; Accepted: 29 June 2019; Published: 3 July 2019

Abstract: In this paper, we propose and evaluate a novel light-emitting diode (LED) nonlinearity estimation and compensation scheme using probabilistic Bayesian learning (PBL) for spectral-efficient visible light communication (VLC) systems. The nonlinear power-current curve of the LED transmitter can be accurately estimated by exploiting PBL regression and hence the adverse effect of LED nonlinearity can be efficiently compensated. Simulation results show that, in a 80-Mbit/s orthogonal frequency division multiplexing (OFDM)-based nonlinear VLC system, comparable bit-error rate (BER) performance can be achieved by the conventional time domain averaging (TDA)-based LED nonlinearity mitigation scheme with totally 20 training symbols (TSs) and the proposed PBL-based scheme with only a single TS. Therefore, compared with the conventional TDA scheme, the proposed PBL-based scheme can substantially reduce the required training overhead and hence greatly improve the overall spectral efficiency of bandlimited VLC systems. It is also shown that the PBL-based LED nonlinearity estimation and compensation scheme is computational efficient for the implementation in practical VLC systems.

Keywords: light emitting diode; nonlinearity estimation and compensation; probabilistic Bayesian learning; visible light communication

1. Introduction

Visible light communication (VLC) relying on white illuminating light-emitting diodes (LEDs) has attracted extensive interest in recent years, due to its inherent advantages such as unregulated spectrum, relatively low implementation cost, enhanced physical-layer security, and electromagnetic interference-free operation [1,2]. The emerging VLC technology has revealed great potential for a lot of practical applications such as high-speed communications, wireless networking, human sensing, ranging and detecting [3,4]. Nevertheless, white LEDs have several limitations which might greatly hinder the development and implementation of VLC systems in practical applications. One limitation is the small modulation bandwidth (typically a few MHz) due to the physical mechanism in the LED quantum well [5] and the long photoluminescence lifetimes of the phosphor, resulting in inter-symbol interference [6]. Several techniques have already been reported for capacity improvement of VLC systems, such as spectral efficiency enhancement employing orthogonal frequency division multiplexing (OFDM) with high-order quadrature amplitude modulation (QAM) constellations [7] and non-orthogonal multiple access [8–11], multiple-input multiple-output (MIMO) transmission [12,13], bandwidth extension using various frequency-domain equalization schemes [14,15], and so on.

Another limitation is that white LEDs suffer from intrinsic nonlinearity which is mainly caused by the thermal effects. It has been shown that LEDs are the major source of nonlinearity in typical VLC systems [16]. Due to LED nonlinearity, the input signal could be severely distorted, especially for OFDM signals which usually have high peak-to-average power ratios (PAPRs) [17]. Generally, there are two approaches to mitigate LED nonlinearity: one is estimation and compensation at transmitter or receiver side and the other is nonlinear equalization. For the first approach, the nonlinear power-current curve of the LED is first estimated, which is then used to compensate the LED nonlinearity. In [18–20], transmitter-side LED nonlinearity estimation and compensation, i.e., pre-distortion, has been considered. In [21,22], receiver-side LED nonlinearity estimation and compensation, i.e., post-distortion, has been applied. For pre-distortion, the estimated nonlinear power-current curve of the LED is treated as a priori information and hence an additional feedback channel is required. In contrast, no feedback is required for post-distortion. For both pre-distortion and post-distortion, time domain averaging (TDA) is usually adopted for accurate estimation of the nonlinear power-current curve of the LED before LED nonlinearity compensation. Nevertheless, conventional TDA usually needs a relatively large number of training symbols (TSs) to achieve the expected performance, which inevitably reduce the spectral efficiency of VLC systems. For the second approach, LED nonlinearity is mitigated by employing various nonlinear equalizers, such as Volterra series-based equalizers [17,23], clustering-based equalizers [24,25], deep learning-based equalizers [26,27], and so on. However, nonlinear equalizers usually require a relatively large training overhead and also suffer from high computational complexity, which might not be suitable for implementation in practical VLC systems due to the limited computing capability of user terminals.

As a widely used machine learning technique, support vector machine (SVM) has already been applied for the mitigation of fiber nonlinearity in coherent optical OFDM systems [28]. However, SVM is a non-probabilistic machine which usually requires a large number of kernels to approximate the optimal solution and hence its application in practical systems is limited. Recently, probabilistic program induction was proposed, which can substantially improve the accuracy of machine learning algorithms when only a few examples are available [29]. In [30], a probabilistic Bayesian learning (PBL) framework was introduced which can obtain a similar generalization performance as that of SVM but needs much fewer basis functions. The PBL technique has many potential applications such as channel estimation [31], radar imagery [32] and frequency-hopping spectrum estimation [33].

In this paper, we for the first time propose a PBL-based LED nonlinearity estimation and compensation scheme for OFDM-based nonlinear VLC systems. The LED nonlinearity, i.e., nonlinear power-current curve, can be accurately estimated by PBL regression and hence the adverse effect of LED nonlinearity can be efficiently compensated at the receiver side. Numerical simulations are performed to validate the feasibility of the proposed PBL-based LED nonlinearity estimation and compensation scheme in a 80-Mbit/s OFDM-based nonlinear VLC system and performance comparison between the proposed PBL-based scheme and the conventional TDA scheme is provided. The computational complexity of the proposed PBL-based LED nonlinearity estimation and compensation scheme is also analyzed.

The rest of this paper is organized as follows. In Section 2, we first introduce the mathematical model of an OFDM-based nonlinear VLC system. Section 3 presents the principle of the proposed PBL-based LED nonlinearity estimation and compensation scheme. The simulation setup is described in Section 4 and the detailed results and discussions are provided in Section 5. Finally, Section 6 concludes the paper.

2. System Model

In this section, we introduce the model of an OFDM-based nonlinear VLC system and the block diagram of the system model is illustrated in Figure 1. As we can see, the input bits are first modulated into real-valued OFDM symbols and then TSs are added for efficient LED nonlinearity estimation and compensation at the receiver side. The obtained digital signal is subsequently transformed into an analog signal through digital-to-analog conversion (DAC), and then a direct current (DC) bias is further added to convert the bipolar signal into a unipolar signal in order to generate a real-valued nonnegative driving signal for the LED transmitter. After that, the generated signal is fed into a white LED which suffers from nonlinearity. In order to support a long transmission distance and a large communication coverage area in typical indoor environments, a relatively high modulation index (MI) is generally required when modulating the signal to the LEDs in practical VLC systems. However, the input signal could be significantly distorted when using a high MI due to LED nonlinearity [34].

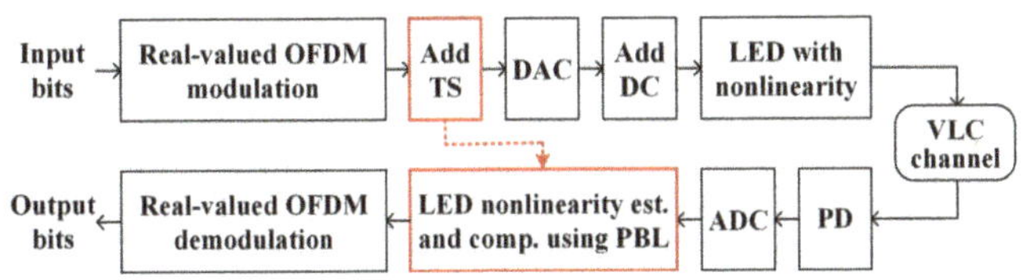

Figure 1. Block diagram of an OFDM-based nonlinear VLC system using PBL-based LED nonlinearity estimation (est.) and compensation (comp.).

The visible light radiated from the white LED propagates through the free-space VLC channel for simultaneous illumination and communication. For simplicity and without loss of generality, it is reasonable to only consider the line-of-sight (LOS) component in the system model [12]. Assuming that the LED follows a generalized Lambertian pattern, the LOS optical channel gain can be calculated by [1]

$$h = \frac{(m+1)\rho A}{2\pi d^2} \cos^m(\varphi) G_f G_l \cos(\theta), \tag{1}$$

where m is the order of Lambertian emission which is given by $m = -\ln 2/\ln(\cos(\Psi))$ with Ψ being the semi-angle at half power of the LED transmitter; ρ and A are the responsivity and the active area of the photodetector (PD), respectively; d is the distance between the LED and the PD; φ and θ are the corresponding emission angle and incident angle, respectively; G_f and G_l are the gains of the optical filter and the optical lens, respectively. The gain of the optical lens is given by $G_l = \frac{n^2}{\sin^2 \Phi}$, where n and Φ are the refractive index and the half-angle field-of-view (FOV) of the optical lens, respectively. Please note that the LOS channel gain becomes zero if the incident light is outside the FOV of the receiver.

At the receiver side, the light is detected by a PD and the obtained electrical OFDM signal can be expressed by [31]

$$y(t) = P_0 h \xi f_x(t) + n(t), \tag{2}$$

where P_0 is the average output optical power of the LED, h is the optical channel gain defined in Equation (1), ξ is the MI of the LED, $f_x(t)$ is the distorted version of the transmitted OFDM signal $x(t)$ due to LED nonlinearity, and $n(t)$ is the additive white Gaussian noise (AWGN) including both shot and thermal noises. The detailed expressions of the noise variances can be found in [31]. The obtained analog OFDM signal is then converted to a digital signal through analog-to-digital conversion (ADC). In order to mitigate the adverse effect of LED nonlinearity in VLC systems, PBL-based LED nonlinearity estimation and compensation are subsequently executed. The detailed procedures of LED nonlinearity estimation and compensation using PBL are described in the next section. After that, the compensated OFDM signal is achieved which is further demodulated to generate the output bits. The principle of real-valued OFDM modulation/demodulation can be found in [31], which is omitted here for brevity.

3. PBL-Based LED Nonlinearity Estimation and Compensation in VLC

3.1. PBL Regression

Following the PBL regression model described in [30,35], for a given data set of input-target pairs $\{s_n, \tau_n\}_{n=1}^{N}$ with length N, the target samples $\{\tau_n\}_{n=1}^{N}$ can be predicted by a linear combination of basis functions:

$$\tau_n = \sum_{m=1}^{M} \mu_m \phi_m(s_n) + \epsilon_n = \boldsymbol{\mu}^T \boldsymbol{\phi}(s_n) + \epsilon_n, \tag{3}$$

where $\boldsymbol{\mu} = [\mu_1, \mu_2, \cdots, \mu_M]^T$ is the parameter vector with length M, $\{\phi_m(s_n)\}_{m=1}^{M}$ is a set of M basis functions and the basis vector is expressed by $\boldsymbol{\phi}(s_n) = [\phi_1(s_n), \phi_2(s_n), \cdots, \phi_M(s_n)]^T$, and $\boldsymbol{\epsilon} = [\epsilon_1, \epsilon_2, \cdots, \epsilon_N]^T$ is the error vector due to the additive noise in the VLC system. Assuming that the error samples $\{\epsilon_n\}_{n=1}^{N}$ are independently and identically distributed Gaussian with zero mean and variance σ^2, a multivariate Gaussian likelihood for the target vector $\boldsymbol{\tau} = [\tau_1, \tau_2, \cdots, \tau_N]^T$ can be written as follows [30]

$$p(\boldsymbol{\tau} \mid \boldsymbol{\mu}, \sigma^2) = \left(2\pi\sigma^2\right)^{-N/2} \exp\left(-\frac{\| \boldsymbol{\tau} - \boldsymbol{\Phi}(s)\boldsymbol{\mu} \|^2}{2\sigma^2}\right), \tag{4}$$

where $\boldsymbol{\Phi}(s) = [\boldsymbol{\phi}(s_1), \boldsymbol{\phi}(s_2), \cdots, \boldsymbol{\phi}(s_N)]^T$ is an $N \times (N+1)$ design matrix with $M = N+1$. According to [30], the basis vector with a bias is defined as $\boldsymbol{\phi}(s_n) = [1, K(s_n, s_1), K(s_n, s_2), \cdots, K(s_n, s_N)]^T$, in which $K(s_i, s_j)$ is the kernel function. In this work, the Gaussian kernel is adopted and hence the kernel function is given by $K(s_i, s_j) = \exp(-\lambda \| s_i - s_j \|^2)$, where λ is known as the width parameter.

From the Bayesian perspective, we can constrain the parameters by defining a zero-mean Gaussian prior distribution over them which takes the form:

$$p(\boldsymbol{\mu} \mid \boldsymbol{\alpha}) = \prod_{n=0}^{N} \mathcal{N}(\mu_n \mid 0, \alpha_n^{-1}), \tag{5}$$

where $\boldsymbol{\alpha} = [\alpha_0, \alpha_1, \alpha_2, \cdots, \alpha_N]^T$ is a vector of $N+1$ independent hyperparameters and each one is used to individually control the strength of the prior over its associated parameter [30]. By combining the likelihood and the prior within Bayes' rule, the posterior parameter distribution conditioned on $\boldsymbol{\tau}$ can be obtained by

$$p(\boldsymbol{\mu} \mid \boldsymbol{\tau}, \boldsymbol{\alpha}, \sigma^2) = \frac{p(\boldsymbol{\tau} \mid \boldsymbol{\mu}, \sigma^2) p(\boldsymbol{\mu} \mid \boldsymbol{\alpha})}{p(\boldsymbol{\tau} \mid \boldsymbol{\alpha}, \sigma^2)}, \tag{6}$$

which is Gaussian $\mathcal{N}(\boldsymbol{w}, \boldsymbol{\Sigma})$ and

$$\boldsymbol{\Sigma} = \left(\sigma^{-2}\boldsymbol{\Phi}^T\boldsymbol{\Phi} + \mathrm{diag}(\boldsymbol{\alpha})\right)^{-1}, \tag{7}$$

$$\boldsymbol{w} = \sigma^{-2}\boldsymbol{\Sigma}\boldsymbol{\Phi}^T\boldsymbol{\tau}. \tag{8}$$

Since it is analytically intractable to include Bayesian inference over those hyperparameters, a type-II maximum likelihood procedure can be used to find a most-probable point estimate $\boldsymbol{\alpha}_{\mathrm{MP}}$ [30]. Therefore, PBL is formulated as the local maximization with respect to $\boldsymbol{\alpha}$ of the marginal likelihood $p(\boldsymbol{\tau} \mid \boldsymbol{\alpha}, \sigma^2)$ and the logarithm of the marginal likelihood is given by [35]

$$\begin{aligned} \log(p(\boldsymbol{\tau} \mid \boldsymbol{\alpha}, \sigma^2)) &= \log \int_{-\infty}^{\infty} p(\boldsymbol{\tau} \mid \boldsymbol{\mu}, \sigma^2) p(\boldsymbol{\mu} \mid \boldsymbol{\alpha}) d\boldsymbol{\mu} \\ &= -\frac{1}{2}\left(N\log 2\pi + \log \mid \boldsymbol{C} \mid + \boldsymbol{\tau}^T \boldsymbol{C}^{-1} \boldsymbol{\tau}\right), \end{aligned} \tag{9}$$

where $C = \sigma^2 I + \Phi(\text{diag}(\alpha))^{-1}\Phi^T$ and I is an identity matrix. Hence, a point estimate w_{MP} for the weights can be obtained by evaluating Equation (8) with $\alpha = \alpha_{\text{MP}}$. As a result, the final prediction of target τ is given by $\hat{\tau} = \Phi(s)w_{\text{MP}}$.

3.2. LED Nonlinearity Estimation and Compensation Using PBL Regression

The principle of the proposed PBL-based LED nonlinearity estimation and compensation scheme is depicted in Figure 2. For simple and efficient LED nonlinearity (i.e., the nonlinear power-current curve of the LED) estimation, a sawtooth-based vector z with N samples, i.e., $z = [z_1, z_2, \cdots, z_N]^T$, is adopted as the TS for LED nonlinearity estimation using PBL regression in this work. The TS is known and shared by all the receivers within the coverage of the VLC system. After transmitting through the nonlinear VLC system, a corresponding vector $r = [r_1, r_2, \cdots, r_N]^T$ can be detected at the receiver side, which can give the raw estimation of the LED nonlinearity. In order to obtain an accurate estimation of the LED nonlinearity, PBL regression is performed which takes the raw estimation of the LED nonlinearity r as the input and the actual LED nonlinearity τ as the target. Although the design matrix is generated by using the input in [30,35], it has been found in our study that the sawtooth-based training vector z can be directly used to generate the design matrix $\Phi(z)$, which can achieve comparable performance as that using the input r. Taking r as the input and using the design matrix $\Phi(z)$, PBL regression can be successfully performed to obtain an accurate estimation of τ, i.e., $\hat{\tau}$.

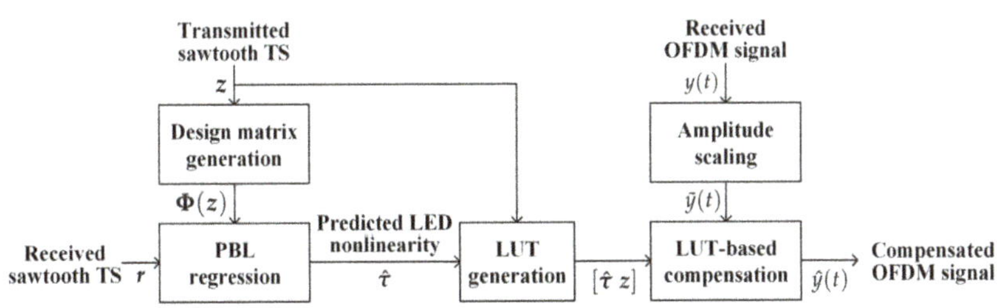

Figure 2. Principle of PBL-based LED nonlinearity estimation and compensation.

As shown in Figure 2, after obtaining z and $\hat{\tau}$, a corresponding look-up table (LUT), i.e., $\mathcal{L} = [\hat{\tau}\ z]$, can be generated. By using the generated LUT, LED nonlinearity compensation can be executed regarding the received OFDM signal. First, the amplitude of the received OFDM signal is scaled by a factor $P_0 h\xi$ so as to match the amplitude of the transmitted OFDM signal, and hence the scaled OFDM signal is expressed by

$$\tilde{y}(t) = \frac{y(t)}{P_0 h\xi} = f_x(t) + \frac{n(t)}{P_0 h\xi}. \tag{10}$$

Then, using the obtained $\hat{\tau} = [\hat{\tau}_1, \hat{\tau}_2, \cdots, \hat{\tau}_N]^T$, the index $i(t)$ of the element in $\hat{\tau}$ which is closest to $\tilde{y}(t)$ can be identified as

$$i(t) = \arg\min_k \{\tilde{y}(t) - \hat{\tau}_k\}, \ k \in \{1, 2, \cdots, N\}, \tag{11}$$

and hence the compensated OFDM signal can be obtained by $\hat{y}(t) = \hat{\tau}_{i(t)}$.

4. Simulation Setup

Numerical simulations using MATLAB are performed to investigate the performance of the proposed PBL-based LED nonlinearity estimation and compensation scheme in an OFDM-based nonlinear VLC system. Key parameters of the simulation setup are listed in Table 1. The LED has a semi-angle at half power of 60° and an output optical power of 10 W. The gain of the optical filter is 0.9. The refractive index and the half-angle FOV of the optical lens are 1.5 and 72°, respectively.

The PD has an active area of 16 mm^2 and a responsivity of 0.53 A/W. The vertical distance between LED and PD is set to 2 m and the horizontal offset between LED and PD is also assumed to be 2 m. A modulation bandwidth of 20 MHz is considered and 16QAM constellation is adopted in real-valued OFDM modulation/demodulation. Hence, the raw data rate of the OFDM-based nonlinear VLC system is 80 Mbit/s. In OFDM modulation/demodulation, the size of fast Fourier transform (FFT)/inverse fast Fourier transform (IFFT) is set to 512. Due to the Hermitian symmetry constraint, only 128 subcarriers are used to carry valid data. A total of 1000 OFDM symbols are transmitted for bit-error rate (BER) calculation.

Table 1. Simulation parameters.

Parameter	Value
Semi-angle at half power of LED	60°
LED output optical power	10 W
Gain of optical filter	0.9
Refractive index of optical lens	1.5
Half-angle FOV of optical lens	72°
Active area of PD	16 mm^2
Responsivity of PD	0.53 A/W
Vertical distance between LED and PD	2 m
Horizontal offset between LED and PD	2 m
Modulation bandwidth	20 MHz
QAM constellation order	16
Raw data rate	80 Mbit/s
Size of FFT/IFFT	512
Number of data subcarriers	128

In the simulation setup, a commercially available white LED (Cree PLCC4) is used and the measured nonlinear power-current curve is illustrated in Figure 3. It can be observed that the normalized output optical power of the LED exhibits strong nonlinearity with the normalized input current. Since only unipolar signal can be modulated onto the luminous intensity, a DC bias current is generally applied to convert the bipolar OFDM signal to a unipolar one. Hence, the nonlinear distortion is mainly caused by two factors: the DC bias current and the peak-to-peak current of the signal. To investigate the LED nonlinearity effect on the transmission performance, as shown in Figure 3, we define the MI as the ratio of the maximum current variation of the signal to the maximum current variation supported by the LED without clipping, and the DC-bias index (DI) as the ratio of the DC bias current of the signal to the maximum current variation supported by the LED without clipping. In this simulation investigation, the LED is assumed to be biased at the middle point of its dynamic range, i.e., DI = 0.5. Although a static LED nonlinearity characteristic is considered in this work, the proposed PBL-based LED nonlinearity estimation and compensation scheme can be easily generalized into an adaptive scheme by adopting the method reported in [36], which is applicable in VLC systems with dynamic LED nonlinearity characteristics.

Moreover, the frequency selective fading effect of a VLC system is mainly caused by the LED, which can be easily compensated by using the frequency domain pre-equalization techniques [14,34]. Thus, a flat system frequency response is considered here without loss of generality.

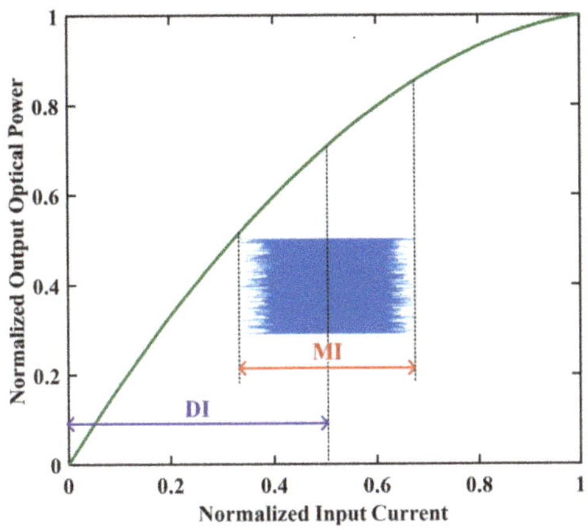

Figure 3. Measured nonlinear power-current curve of white LED Cree PLCC4.

The original PBL regression model proposed in [30] starts with all the $M = N + 1$ basis functions. The update rules for the hyperparameters depend on computing the posterior weight co-variance matrix which requires a Cholesky decomposition operation with a complexity of $O(M^3)$. As a result, the computational complexity of PBL regression could be very high for practical applications. To reduce computational complexity, a fast marginal likelihood maximization method has been proposed in [35]. Based on an accelerated training algorithm, PBL regression is initialized with a single basis function, i.e., the bias. Sequentially, the basis functions are added to increase the marginal likelihood and also modify their weightings, and meanwhile the redundant basis functions are deleted to increase the objective function. Hence, the new PBL regression model can achieve comparable performance as the original one but with greatly reduced computational complexity, which is adopted here to realize efficient LED nonlinearity estimation and compensation in OFDM-based nonlinear VLC systems. Specifically, the width parameter of the Gaussian kernel is set to $\lambda = 2$ and the maximum number of iterations is set to 10. For the purpose of comparison, the conventional TDA scheme employing multiple consecutive TSs is also considered in the simulations.

5. Results and Discussion

In this section, simulation results based on the setup described above are presented and the corresponding discussions are also provided. Figure 4 depicts the normalized signal amplitude vs. the normalized input current for different MI values, where DI = 0.5 and the length of the sawtooth-based TS is $N = 64$. As we can observe, for both MI = 0.6 and 0.8, the amplitude of the received TS has a significant variation due to the additive noise in the VLC system. However, after performing PBL regression, the predicted power-current curve of the LED becomes much smoother which matches the actual power-current curve of the LED very well, indicating an accurate estimation of the LED nonlinearity by using only a single sawtooth-based TS.

We further analyze the impact of the number of TSs on the estimation accuracy of both the conventional TDA scheme and the proposed PBL-based scheme. Figure 5 shows the average estimation error vs. the number of TSs with MI = 0.8, DI = 0.5 and $N = 64$. For the conventional TDA scheme, it can be seen that the average estimation error is gradually reduced with the increase of the number of TSs and it becomes stable at about 1.7×10^{-3} when the number of TSs reaches 20. In contrast, for the proposed PBL regression scheme, a comparable average estimation error can be achieved by using only a single TS.

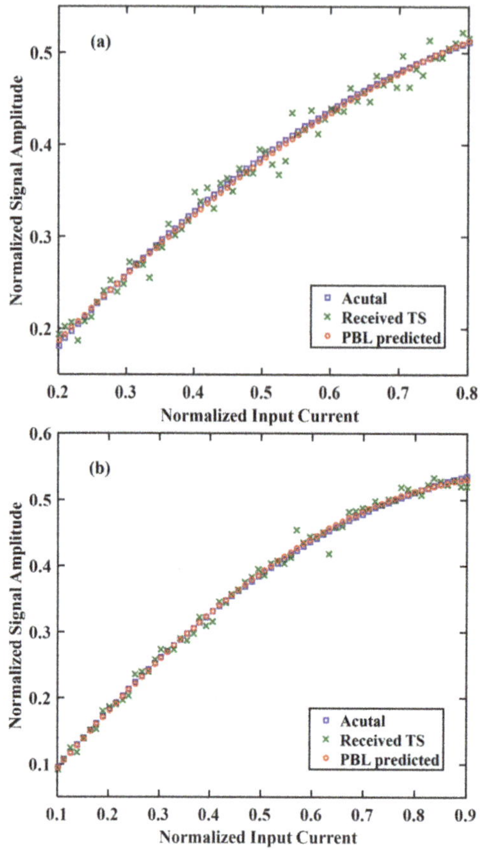

Figure 4. Normalized signal amplitude vs. normalized input current for (**a**) MI = 0.6 and (**b**) MI = 0.8.

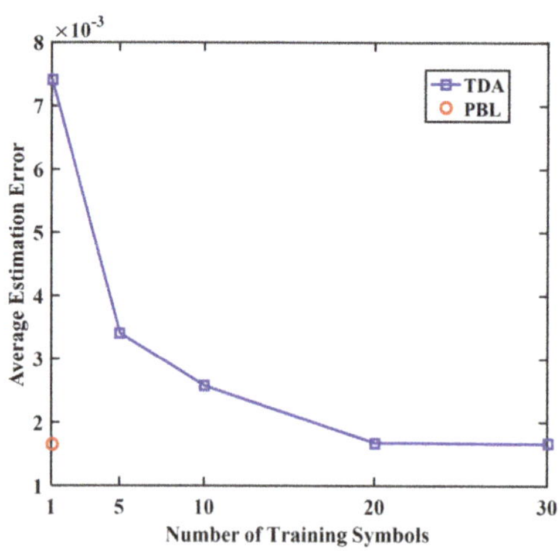

Figure 5. Average estimation error vs. number of training symbols.

The impact of the length of the TS on the BER performance of the OFDM-based nonlinear VLC system is also analyzed. Figure 6 shows the BER vs. the length of the TS for the conventional

TDA scheme with different numbers of TSs and the proposed PBL-based scheme with a single TS, with MI = 0.8 and DI = 0.5. Evidently, for both conventional TDA and the proposed PBL-based scheme, the BER performance is substantially improved with the increase of the length of the TS and stable BERs can be guaranteed when the length of the TS is about 64.

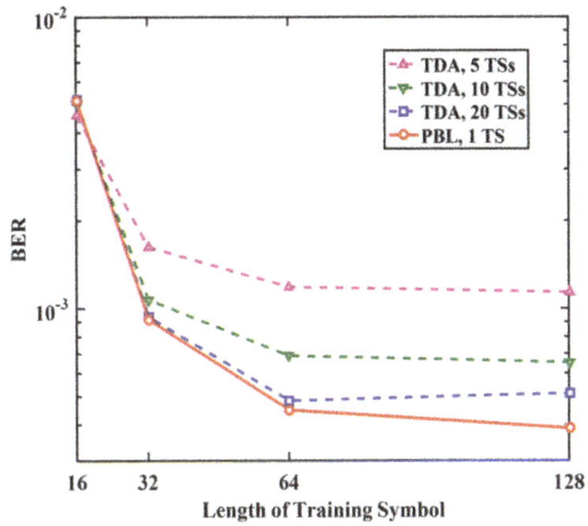

Figure 6. BER vs. length of training symbol.

Based on the analysis above, we evaluate and compare the BER performance vs. MI for the OFDM-based nonlinear VLC system without and with LED nonlinearity compensation, where DI = 0.5 and the length of the TS is fixed at $N = 64$. As shown in Figure 7, when no LED nonlinearity compensation is considered, the BER reduces with the increase of the MI and an MI of 0.87 is required to reach the BER threshold of 10^{-3}. Moreover, the BER improvement becomes insignificant by further increasing MI when MI is larger than 0.9, suggesting the adverse effect of LED nonlinearity on the BER performance of the OFDM-based nonlinear VLC system. However, when the conventional TDA scheme is applied, the BER performance can be substantially improved. For TDA with 5 TSs, the required MI to reach BER = 10^{-3} is reduced to 0.80. By further increasing the number of TSs, the BER performance can be further enhanced. In contrast, when the proposed PBL-based scheme with a single TS is adopted, the required MI to reach BER = 10^{-3} is only about 0.77. Furthermore, nearly the same BER performance can be achieved by the conventional TDA scheme with totally 20 TSs and the proposed PBL-based scheme with a single TS. Therefore, a substantial reduction of the training overhead for LED nonlinearity mitigation can be achieved by using the proposed PBL-based scheme in comparison to the conventional TDA scheme, which indicates a greatly improved overall spectral efficiency of the OFDM-based nonlinear VLC system. The corresponding 16QAM constellation diagrams are also shown as insets in Figure 7.

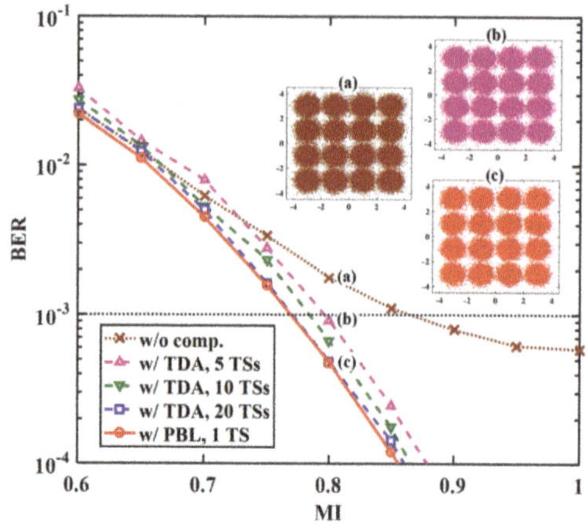

Figure 7. BER vs. MI without and with LED nonlinearity compensation.

Finally, we analyze the computational complexity of the proposed PBL-based LED nonlinearity estimation and compensation scheme. For the fast PBL regression model adopted here, PBL regression is initialized with the bias only and the basis functions are iteratively added, updated or deleted to increase the marginal likelihood [35]. Since adding basis functions requires most of the computations, the worst case is that a new basis function is added at each iteration and the worst-case computational complexity is given by $O(N_{it}M^2)$ with N_{it} being the iteration number and M being the number of basis functions. However, the worst case scenarios are highly impossible to occur due to the inherent sparsity of PBL regression. As introduced in [31], for common cases, an approximation of the computational complexity is about $O(N_{it}N_{nz}^2)$ with N_{nz} being the number of non-zero elements in the weight vector w_{MP}. More specifically, by using the fast marginal likelihood maximization method, only three elements out of totally $M = 65$ elements in w_{MP} are non-zero, i.e., $N_{nz} = 3$. Consequently, the proposed PBL-based LED nonlinearity estimation and compensation scheme is computational efficient which is suitable for the implementation in practical VLC systems with computing capability-limited user terminals.

6. Conclusions

In this paper, we have proposed a novel LED nonlinearity estimation and compensation scheme based on PBL regression for OFDM-based nonlinear VLC systems. The performance of the proposed PBL-based scheme has been evaluated by numerical simulations and further compared with the conventional TDA scheme. The obtained simulation results have shown that the proposed PBL-based scheme can accurately estimate the nonlinear power-current curve of the LED and hence efficiently compensate the adverse effect of LED nonlinearity. More specifically, the proposed PBL-based scheme with a single TS can achieve a comparable BER performance as the conventional TDA scheme with totally 20 TSs. Therefore, the required training overhead for LED nonlinearity mitigation can be substantially reduced and the overall system spectral efficiency can be greatly improved by adopting the proposed PBL-based scheme. It is also shown that the proposed PBL-based LED nonlinearity estimation and compensation scheme is computational efficient, which is suitable for potential application in practical VLC systems.

Author Contributions: Conceptualization, C.C., X.D., Y.Y. and L.Z.; Formal analysis, C.C., X.D., Y.Y. and H.Y.; Funding acquisition, C.C.; Project administration, C.C. and P.D.; Writing—original draft, C.C. and P.D.; Writing—review & editing, C.C., X.D., Y.Y., P.D., H.Y. and L.Z.

Funding: This research was funded by the Starting Research Fund from the Chongqing University (No. 02140011044110).

Conflicts of Interest: The authors declare no conflict of interest.

References

1. Komine, T.; Nakagawa, M. Fundamental analysis for visible-light communication system using LED lights. *IEEE Trans. Consum. Electron.* **2004**, *50*, 100–107. [CrossRef]
2. Jovicic, A.; Li, J.; Richardson, T. Visible light communication: Opportunities, challenges and the path to market. *IEEE Commun. Mag.* **2013**, *51*, 26–32. [CrossRef]
3. Pathak, P.H.; Feng, X.; Hu, P.; Mohapatra, P. Visible light communication, networking, and sensing: A survey, potential and challenges. *IEEE Commun. Surv. Tutor.* **2015**, *17*, 2047–2077. [CrossRef]
4. Wu, D.; Zhong, W.D.; Ghassemlooy, Z.; Chen, C. Short-range visible light ranging and detecting system using illumination light emitting diodes. *IET Optoelectron.* **2016**, *10*, 94–99. [CrossRef]
5. Deng, X.; Mardanikorani, S.; Wu, Y.; Arulandu, K.; Chen, B.; Khalid, A.M.; Linnartz, J.P.M. Mitigating LED nonlinearity to enhance visible light communications. *IEEE Trans. Commun.* **2018**, *66*, 5593–5607. [CrossRef]
6. Rajbhandari, S.; Chun, H.; Faulkner, G.; Cameron, K.; Jalajakumari, A.V.; Henderson, R.; Tsonev, D.; Ijaz, M.; Chen, Z.; Haas, H.; et al. High-speed integrated visible light communication system: Device constraints and design considerations. *IEEE J. Sel. Areas Commun.* **2015**, *33*, 1750–1757. [CrossRef]
7. Elgala, H.; Mesleh, R.; Haas, H.; Pricope, B. OFDM visible light wireless communication based on white LEDs. In Proceedings of the IEEE 65th Vehicular Technology Conference (VTC Spring), Dublin, Ireland, 22–25 April 2007; pp. 2185–2189.
8. Marshoud, H.; Kapinas, V.M.; Karagiannidis, G.K.; Muhaidat, S. Non-orthogonal multiple access for visible light communications. *IEEE Photonics Technol. Lett.* **2015**, *28*, 51–54. [CrossRef]
9. Chen, C.; Zhong, W.D.; Yang, H.; Du, P. On the performance of MIMO-NOMA-based visible light communication systems. *IEEE Photonics Technol. Lett.* **2017**, *30*, 307–310. [CrossRef]
10. Yang, Y.; Chen, C.; Zhang, W.; Deng, X.; Du, P.; Yang, H.; Zhong, W.D.; Chen, L. Secure and private NOMA VLC using OFDM with two-level chaotic encryption. *Opt. Express* **2018**, *26*, 34031–34042. [CrossRef]
11. Chen, C.; Zhong, W.D.; Yang, H.; Du, P.; Yang, Y. Flexible-rate SIC-free NOMA for downlink VLC based on constellation partitioning coding. *IEEE Wirel. Commun. Lett.* **2019**, *8*, 568–571. [CrossRef]
12. Zeng, L.; O'Brien, D.C.; Le Minh, H.; Faulkner, G.E.; Lee, K.; Jung, D.; Oh, Y.; Won, E.T. High data rate multiple input multiple output (MIMO) optical wireless communications using white LED lighting. *IEEE J. Sel. Areas Commun.* **2009**, *27*, 1654–1662. [CrossRef]
13. Chen, C.; Zhong, W.D.; Wu, D. On the coverage of multiple-input multiple-output visible light communications. *IEEE/OSA J. Opt. Commun. Netw.* **2017**, *9*, D31–D41. [CrossRef]
14. Chen, C.; Zhong, W.D.; Wu, D. Indoor OFDM visible light communications employing adaptive digital pre-frequency domain equalization. In Proceedings of the Conference on Lasers and Electro-Optics (CLEO), San Jose, CA, USA, 5–10 June 2016; paper JTh2A.118.
15. Le Minh, H.; O'Brien, D.; Faulkner, G.; Zeng, L.; Lee, K.; Jung, D.; Oh, Y.; Won, E.T. 100-Mb/s NRZ visible light communications using a postequalized white LED. *IEEE Photonics Technol. Lett.* **2009**, *21*, 1063–1065. [CrossRef]
16. Ying, K.; Yu, Z.; Baxley, R.J.; Qian, H.; Chang, G.K.; Zhou, G.T. Nonlinear distortion mitigation in visible light communications. *IEEE Wirel. Commun.* **2015**, *22*, 36–45. [CrossRef]
17. Stepniak, G.; Siuzdak, J.; Zwierko, P. Compensation of a VLC phosphorescent white LED nonlinearity by means of Volterra DFE. *IEEE Photonics Technol. Lett.* **2013**, *25*, 1597–1600. [CrossRef]
18. Mitra, R.; Bhatia, V. Chebyshev polynomial-based adaptive predistorter for nonlinear LED compensation in VLC. *IEEE Photonics Technol. Lett.* **2016**, *28*, 1053–1056. [CrossRef]
19. Liang, S.; Jiang, Z.; Qiao, L.; Lu, X.; Chi, N. Faster-than-Nyquist precoded CAP modulation visible light communication system based on nonlinear weighted look-up table predistortion. *IEEE Photonics J.* **2018**, *10*, 1–9. [CrossRef]
20. Lu, H.; Jin, J.; Wang, J. Alleviation of LED nonlinearity impact in visible light communication using companding and predistortion. *IET Commun.* **2019**, *13*, 818–821. [CrossRef]
21. Mitra, R.; Bhatia, V. Adaptive sparse dictionary-based kernel minimum symbol error rate post-distortion for nonlinear LEDs in visible light communications. *IEEE Photonics J.* **2016**, *8*, 1–13. [CrossRef]

22. Deng, X.; Mardanikorani, S.; Arulandu, K.; Linnartz, J.P.M. Novel Post-distortion to Mitigate LED Nonlinearity in High-speed Visible Light Communications. In Proceedings of the IEEE Globecom Workshops (GC Wkshps), Abu Dhabi, United Arab Emirates, 9–13 December 2018; pp. 1–6.
23. Wang, Y.; Tao, L.; Huang, X.; Shi, J.; Chi, N. Enhanced performance of a high-speed WDM CAP64 VLC system employing Volterra series-based nonlinear equalizer. *IEEE Photonics J.* **2015**, *7*, 1–7. [CrossRef]
24. Lu, X.; Wang, K.; Qiao, L.; Zhou, W.; Wang, Y.; Chi, N. Nonlinear compensation of multi-CAP VLC system employing clustering algorithm based perception decision. *IEEE Photonics J.* **2017**, *9*, 1–9. [CrossRef]
25. Ma, J.; He, J.; Shi, J.; Zhou, Z.; Deng, R. Nonlinear Compensation Based on K-Means Clustering Algorithm for Nyquist PAM-4 VLC System. *IEEE Photonics Technol. Lett.* **2019**, *31*, 935–938. [CrossRef]
26. Lu, X.; Lu, C.; Yu, W.; Qiao, L.; Liang, S.; Lau, A.P.T.; Chi, N. Memory-controlled deep LSTM neural network post-equalizer used in high-speed PAM VLC system. *Opt. Express* **2019**, *27*, 7822–7833. [CrossRef] [PubMed]
27. Miao, P.; Zhu, B.; Qi, C.; Jin, Y.; Lin, C. A Model-Driven Deep Learning Method for LED Nonlinearity Mitigation in OFDM-based Optical Communications. *IEEE Access* **2019**, *7*, 71436–71446. [CrossRef]
28. Nguyen, T.; Mhatli, S.; Giacoumidis, E.; Van Compernolle, L.; Wuilpart, M.; Mégret, P. Fiber nonlinearity equalizer based on support vector classification for coherent optical OFDM. *IEEE Photonics J.* **2016**, *8*, 1–9. [CrossRef]
29. Lake, B.M.; Salakhutdinov, R.; Tenenbaum, J.B. Human-level concept learning through probabilistic program induction. *Science* **2015**, *350*, 1332–1338. [CrossRef]
30. Tipping, M.E. Sparse Bayesian learning and the relevance vector machine. *J. Mach. Learn. Res.* **2001**, *1*, 211–244.
31. Chen, C.; Zhong, W.D.; Zhao, L. Sparse Bayesian RVM regression based channel estimation for IM/DD OFDM-VLC systems with reduced training overhead. In Proceedings of the IEEE International Conference on Communications Workshops (ICC Workshops), Paris, France, 21–25 May 2017; pp. 162–167.
32. Zhao, L.; Wang, L.; Yang, L.; Zoubir, A.M.; Bi, G. The race to improve radar imagery: An overview of recent progress in statistical sparsity-based techniques. *IEEE Signal Process. Mag.* **2016**, *33*, 85–102. [CrossRef]
33. Zhao, L.; Wang, L.; Bi, G.; Zhang, L.; Zhang, H. Robust frequency-hopping spectrum estimation based on sparse Bayesian method. *IEEE Trans. Wirel. Commun.* **2014**, *14*, 781–793. [CrossRef]
34. Li, J.; Huang, Z.; Liu, X.; Ji, Y. Hybrid time-frequency domain equalization for LED nonlinearity mitigation in OFDM-based VLC systems. *Opt. Express* **2015**, *23*, 611–619. [CrossRef]
35. Tipping, M.E.; Faul, A.C. Fast marginal likelihood maximisation for sparse Bayesian models. In Proceedings of the 9th International Workshop on Artificial Intelligence and Statistics, Key West, FL, USA, 3–6 January 2003; pp. 1–13.
36. Aggarwal, P.; Kabra, T.; Ahmad, R.; Bohara, V.A.; Srivastava, A. Adaptive learning architecture-based predistorter for nonlinear VLC system. *Photonics Netw. Commun.* **2019**, 1–12. [CrossRef]

applied sciences

MDPI

Article

Deep Neural Network Equalization for Optical Short Reach Communication

Maximilian Schaedler [1,2,*], Christian Bluemm [1], Maxim Kuschnerov [1], Fabio Pittalà [1], Stefano Calabrò [1] and Stephan Pachnicke [2]

1 Huawei Technologies, German Research Center, Riesstr. 25, 80992 Munich, Germany; christian.bluemm@huawei.com (C.B.); maxim.kuschnerov@huawei.com (M.K.); fabio.pittala@huawei.com (F.P.); stefano.calabro@huawei.com (S.C.)

2 Chair of Communications, Institute of Electrical Engineering and Information Technology, Kiel University, Kaiserstr. 2, 24143 Kiel, Germany; stephan.pachnicke@tf.uni-kiel.de

* Correspondence: maximilian.schaedler@stu.uni-kiel.de or maximilian.schaedler@huawei.com

Received: 13 September 2019; Accepted: 29 October 2019; Published: 2 November 2019

Abstract: Nonlinear distortion has always been a challenge for optical communication due to the nonlinear transfer characteristics of the fiber itself. The next frontier for optical communication is a second type of nonlinearities, which results from optical and electrical components. They become the dominant nonlinearity for shorter reaches. The highest data rates cannot be achieved without effective compensation. A classical countermeasure is receiver-side equalization of nonlinear impairments and memory effects using Volterra series. However, such Volterra equalizers are architecturally complex and their parametrization can be numerical unstable. This contribution proposes an alternative nonlinear equalizer architecture based on machine learning. Its performance is evaluated experimentally on coherent 88 Gbaud dual polarization 16QAM 600 Gb/s back-to-back measurements. The proposed equalizers outperform Volterra and memory polynomial Volterra equalizers up to 6th orders at a target bit-error rate (BER) of 10^{-2} by 0.5 dB and 0.8 dB in optical signal-to-noise ratio (OSNR), respectively.

Keywords: deep neural networks; volterra equalization; nonlinear systems; coherent optical communication

1. Introduction

Modern communication networks build upon a backbone of optical systems, which have fiber as the transmission medium. Due to its physical properties, including the Kerr effect, fiber communication is always affected by nonlinearities. This is the obvious, but not the only major source of nonlinear distortion in optical communication systems. Regardless of the individual technology choices for the optical transceiver architecture, it will always comprise a large number of optical/electrical (O/E) components with nonlinear transfer characteristics, often with memory effects. Their limitations on the achievable capacity are aggravated by measures towards higher data rates, such as symbol rate increase or the shift to higher order pulse amplitude modulations (PAM) and quadrature amplitude modulations (QAM). Nonlinear compensation for O/E components is not yet a standard feature in today's optical transceivers, but will inevitably become a key element of digital signal processing (DSP) to keep up with ever increasing data rates. Especially when high-bandwidth (BW) communication meets short reach, reduced fiber lengths turn components into the dominant source of nonlinearities. Typical use cases include data center interconnect (DCI) with a range of 80–120 km [1]. Volterra nonlinear equalizers (VNLE) have proven very effective against both, fiber nonlinearities [2] and component nonlinearities [3,4]. They approximate with analytical models based on Volterra series, which can be tailored to match any nonlinear system by choosing a high enough polynomial order

P and memory depth M [5] and a set of coefficients (kernels), one for each order/memory tap combination. Excessive scalability, however, can become a crucial downside, as the architectural complexity of VNLEs increases exponentially with P and M. A popular reduced-size alternative are memory polynomial (MP) VNLEs, which operate only on a subset of kernels [6]. However, this reduction costs effectiveness.

Another critical disadvantage of both VNLEs and MP-VNLEs is numerical instability when identifying and extracting the VNLE kernels with the preferred approach, a least squares (LS) solver. This limits in practice the maximum polynomial order P and memory depth M. These shortcomings make general VNLEs and MP-VNLEs far from ideal, which motivate to assess an alternative equalizer structure based on deep neural networks (DNN) in this article. The performance of VNLEs and MP-VNLEs is benchmarked against DNN nonlinear equalizers (NLE) on basis of coherent dual polarization (DP) 16QAM back-to-back (BtB) offline captures, which are reprocessed for all options. For the measurements, a DCI compatible net bit rate of 600 Gb/s with 15% FEC overhead has been chosen. We like to highlight that the universal and flexible nature of both Volterra-based equalizers and DNN equalizers is not limited to this exemplary configuration. One can expect benefits for different reaches, data rates or even detection technologies, such as IM-DD with PAM modulation. We focus on coherent systems here, since we target a single lambda system for our 600 Gb/s data rate. This is out of scope for IM-DD, due to its limited spectral efficiency.

2. Principles of Nonlinear Equalizer

The presented BtB setup focuses on component nonlinearities, while the compensation of fiber nonlinearities has been shown before [7]. For this paper, DNN structures have been implemented from scratch with Matlab, not using any machine learning (ML) toolbox. Being trained on the same data as (MP-)VNLEs, DNNs learn to capture nonlinear effects of O/E components and fiber, as well as their memory characteristics.

2.1. General Volterra Equalizer

With $y(n)$ and $\tilde{y}(n)$ representing system input and output, respectively, the Pth-order discrete time Volterra series with M_p memory taps for order p is given as [8]

$$\tilde{y}(n) = \sum_{p=1}^{P} \sum_{m_1=0}^{M_1} \cdots \sum_{m_p=0}^{M_p} h_p(m_1, \cdots, m_p) \prod_{k=1}^{p} y(n - m_k) \tag{1}$$

It is the most complete model for nonlinearities with memory, as the pth-order Volterra kernel $h_p(m_1, \cdots, m_p)$ includes *all* possible combinations of a product of p time shifts of the input signal up to memory depth m_p.

Figure 1 illustrates this mapping of memory and nonlinearity in respect to the input signal $y(n)$, which can be either real- or complex-valued. A VNLE architecture with given memory size and order is fully described by its so-called kernels $h_p(m_1, \cdots, m_p)$. Before operation, they have to be identified, i.e., configured upon training data, in order to match the nonlinearities of interest. To this means, iterative approaches exist, such as least mean squares (LMS) or recursive least squares (RLS) kernel extraction. They are good solutions if little training data is available at a time or if channel dynamics call for frequent kernel reconfiguration. Typical O/E channels, however, are static enough to allow for a one shot configuration on a sufficiently large data set. The optimal solution in terms of the least squares (LS) error criterion to match the equalizer output to the desired signal (training signal) is achieved with a single operation LS solver. It involves inverting matrices, built from transmitted and received training data [9]. It has been shown [10] that these matrices can be ill-conditioned, resulting in a huge ratio between the highest and lowest eigenvalues. As a result, this matrix becomes nearly singular. With growing matrix complexity, i.e., increasing P or M, the inverse of the matrix is computed with less and less accuracy. This calls for a trade off between matrix conditioning and the number of kernels for maximum performance.

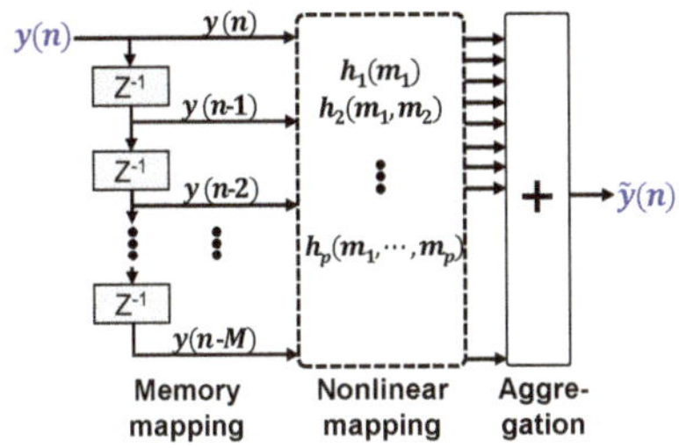

Figure 1. General nonlinear volterra equalizer.

2.2. Memory Polynomial Volterra Equalizer

Shrinking the architectural complexity of VNLEs makes sense for two reasons. Less digital signal processing (DSP) resources are needed for deployment in actual transceivers and the LS matrices become less ill-conditioned. Figure 2 illustrates the MP-VNLE structure. MP-VNLEs prune general VNLEs simply by removing all cross terms generated by $y(n)$ and its delayed versions $y(n-m_k)$:

$$\tilde{y}(n) = \sum_{p=1}^{P}\sum_{m=0}^{M} h_{pm}y(n-m)|y(n-m)|^{p-1} \tag{2}$$

One may rewrite this as

$$\tilde{y}(n) = \sum_{m=0}^{M} g_m y(n-m) \quad \text{with } g_m = \sum_{p=1}^{P} h_{pm}|y(n-m)|^{p-1} \tag{3}$$

Given a memory tap m, g_m becomes a constant gain to the memory term $y(n-m)$. g_m, in turn, depends only on the polynomial power terms of $|y(n)|$. Again, $y(n)$ can be real- or complex-valued.

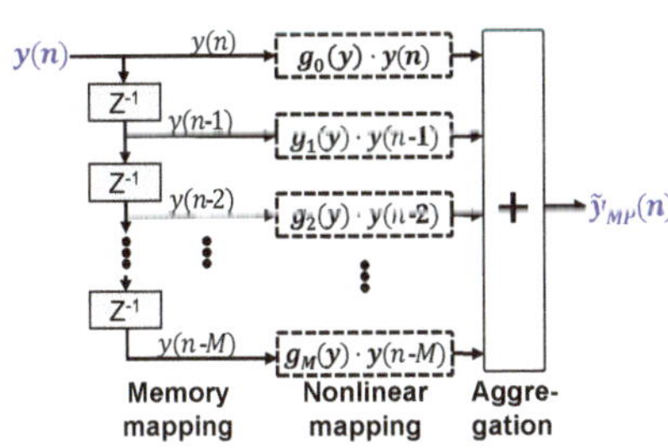

Figure 2. Memory polynomial nonlinear equalizer.

2.3. Deep Neural Network Equalizer

Along with other science and engineering disciplines, the optical communication community is adopting ML for a broad range of problems [11–13]. Neural network (NN) is a subcategory of ML, which automatically learns systematic features within an arbitrary data set. It then can extract and extrapolate these features to new data.

NNs computing structures are built from several layers of artificial neurons. A single artificial neuron is a processing unit with a number of inputs and one output. Each input is associated with a weight. The neuron firstly computes an activation by summing up the particular weighted inputs and a bias term. Secondly, an activation function $\sigma(\cdot)$ is applied to obtain the neurons's output $a = \sigma(z)$. The neurons's behaviour is defined by the weights, bias and the activation function [14]. Interconnecting multiple neurons builds a NN. If more than one hidden layer is used, the NN is called a deep NN, referred to as DNN. DNNs are convenient to model unknown highly complex relations. As such, they are capable of learning nonlinear equalization. In machine learning terms, nonlinear equalization can be treated as a problem of supervised nonlinear regression. A regression DNN learns a function $f(\cdot) : \mathcal{R}^M \rightarrow \mathcal{R}^A$ by training on a dataset, where M is the input dimension and A is the output dimension. Figure 3 shows a basic architecture of a regressive neural network with multiple hidden layers. The bias term is included by an additional branch to each neuron. The notation is as follows [14] :

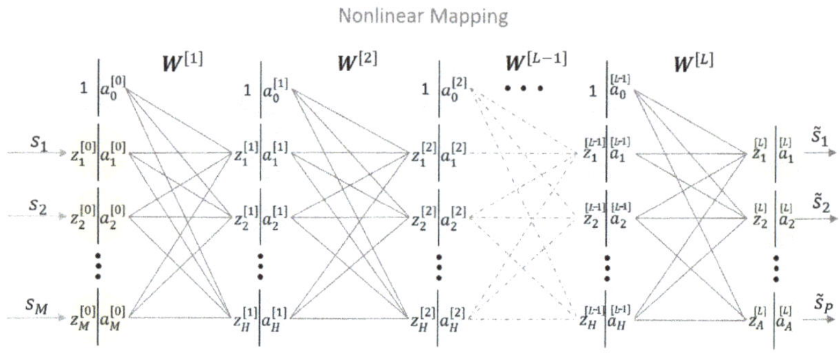

Figure 3. L-layers feed-forward deep neural networks (DNN) with M nodes in the input layer, H nodes in the hidden layers and A nodes in the output layer.

- $\underline{s} = [s_1, ..., s_M]$ denotes the input signal,
- $\underline{\tilde{s}} = [\tilde{s}_1, ..., \tilde{s}_A]$ denotes the equalized signal,
- $\mathbf{W}^{[l]}$ denotes a weight matrix (including the bias) between the $(l-1)$-th layer to the l-th layer,
- $a_i^{[l]}$ denotes the output of the i-th neuron in the l-th layer,
- $z_i^{[l]}$ denotes the summed up input to the i-th neuron in the l-th layer,
- The activation function $\sigma_i^{[l]}$ is the relation between $z_i^{[l]}$ and $a_i^{[l]}$ of the i-th neuron in the l-th layer,

where l denotes the current layer and H the number of nodes in the hidden layers. Using these notations the output can be expressed as follows [14],

$$\underline{a}^{[0]} = \underline{s}, \tag{4}$$

$$\underline{z}^{[l]} = \mathbf{W}^{[l]} \underline{a}^{[l-1]}, \tag{5}$$

$$\underline{a}^{[l]} = \sigma^{[l]}\left(\underline{z}^{[l]}\right), \tag{6}$$

where the first line, Equation (4), denotes the input layer and the second as well as the third lines, Equations (5) and (6), are executed iteratively for subsequent layers to obtain the output $\underline{a}^{[L]}$. For regression problems, the activation function in the last layer may differ from the activation functions used in the hidden layers. The hidden layers usually use nonlinear functions, while the type of activation function at the last layer should be matched to the application. This output activation function has to represent the overall dynamic range of the target signal. A common choice is therefore the identity activation function for a regression problem. The identity activation function is applied to each neuron at the output layer. The DNN models the relation of $\underline{s}$ and $\underline{\tilde{s}}$ as being an output of a DNN with respect to the input $\underline{s}$ and the nonlinear functions. The equalized signal is obtained at the output of the DNN and can be expressed as follows,

$$\underline{\tilde{s}}_{\text{DNN-NLE}} = \underline{a}^{[L]}(\underline{s}, \mathbf{W}). \tag{7}$$

In order to find the weights such that the outputs $\underline{a}^{[L]}$ are close to the target outputs $\underline{\tilde{s}}$ and to assess performance of the DNN, a cost function $C(\underline{s}, \underline{c}, \mathbf{W})$ has to be introduced. The cost function is derived using the Maximum Likelihood principle. Considering the regression problem with the training data set of N inputs $\{\underline{s}_1, ..., \underline{s}_N\}$, along with the corresponding target output values $\{\underline{\tilde{s}}_1, ..., \underline{\tilde{s}}_N\}$,

$$\mathcal{D}_{\text{Dataset}} = \left\{ (\underline{s}_i, \underline{\tilde{s}}_i) : \underline{s}_i \in \mathcal{R}^M, \underline{\tilde{s}}_i \in \mathcal{R}^A, \quad i = 1, ..., N \right\}, \tag{8}$$

where each $\underline{s}_i$ has M entries and each $\underline{\tilde{s}}_i$ has A entries for all $i = 1, ..., N$. The error between the DNN output and the target values is defined as

$$\underline{\epsilon}_i = \underline{\tilde{s}}_i - \underline{a}^{[L]}(\underline{s}_i, \mathbf{W}^{[1]}, ..., \mathbf{W}^{[L]}). \tag{9}$$

Using Equation (9) and assuming a Gaussian distribution of the error with zero mean and variance σ^2, as well as statistically independent symbols, the joint conditional probability density function can be derived,

$$p(\tilde{s}_1, ..., \tilde{s}_N | s_1, ..., s_N, \mathbf{W}) = \prod_{i=1}^{N} \frac{1}{\sqrt{2\pi\sigma^2}} exp\left(- \frac{(\underline{\tilde{s}}_i - \underline{a}^{[L]}(\underline{s}_i, \mathbf{W}^{[1]}, ..., \mathbf{W}^{[L]}))^2}{2\sigma^2} \right). \tag{10}$$

According to the ML principle, our target is to find the Maximum Likelihood estimates of **W** for the DNN which gives the highest probability of our training data. This can be done by maximizing the joint conditional probability density function Equation (10).

$$\begin{aligned}
(\mathbf{W}_{\text{ML}}^{[1]}, ..., \mathbf{W}_{\text{ML}}^{[L]}) &= \underset{\mathbf{W}_{\text{ML}}^{[l]}, l=1,..,L}{\arg\max} \; p_{\underline{\tilde{s}}^N | \underline{s}^N}(\tilde{s}_1, ..., \tilde{s}_N | s_1, ..., s_N, \mathbf{W}^{[l]}) && (11) \\
&= \underset{\mathbf{W}_{\text{ML}}^{[l]}, l=1,..,L}{\arg\max} \prod_{i=1}^{N} \frac{1}{\sqrt{2\pi\sigma^2}} exp\left(- \frac{(\underline{\tilde{s}}_i - \underline{a}^{[L]}(\underline{s}_i, \mathbf{W}^{[l]}))^2}{2\sigma^2} \right) && (12) \\
&= \underset{\mathbf{W}_{\text{ML}}^{[l]},, l=1,..,L}{\arg\min} \sum_{i=1}^{N} \frac{1}{2} \left(\underline{\tilde{s}}_i - \underline{a}^{[L]}(\underline{s}_i, \mathbf{W}^{[l]}) \right)^2 && (13) \\
&= \underset{\mathbf{W}_{\text{ML}}^{[l]}, l=1,..,L}{\arg\min} \; C(\mathbf{W}^{[l]}) && (14)
\end{aligned}$$

By applying the logarithm on Equation (12), we obtain the quadratic cost function. Thus, minimizing the quadratic cost function corresponds to finding the ML estimate of weights for the DNN. Equation (14) defines the final overall cost of all samples in the dataset. In order to find the weights, which minimize the quadratic cost function, the gradient descent algorithm [15] is used. Here, the gradients of the cost function have to be computed with respect to the DNN parameters. An efficient method is the common backpropagation algorithm [16]. The backpropagation algorithm approaches the derivatives of the cost function from the training set with respect to the network parameters. It applies the chain rule for derivatives very efficiently. The derivation of the cost function with respect to the weights in the Lth-layer down to the first layer by applying the chain rule, can be expressed as follows

$$\frac{\partial C(\mathbf{W}^{[L]})}{\partial \mathbf{W}^{[L]}} = \frac{\partial C(\underline{a}^{[L]})}{\partial \underline{a}^{[L]}} \frac{\partial \underline{a}^{[L]}(\underline{z}^{[L]})}{\partial \underline{z}^{[L]}} \frac{\partial \underline{z}^{[L]}(\mathbf{W}^{[L]})}{\partial \mathbf{W}^{[L]}}, \tag{15}$$

$$\frac{\partial C(\mathbf{W}^{[l]})}{\partial \mathbf{W}^{[l]}} = \frac{\partial C(\underline{z}^{[l+1]})}{\partial \underline{z}^{[l+1]}} \frac{\partial \underline{z}^{[l+1]}(\underline{a}^{[l]})}{\partial \underline{a}^{[L]}} \frac{\partial \underline{a}^{[l]}(\underline{z}^{[l]})}{\partial \underline{z}^{[L]}} \frac{\partial \underline{z}^{[l]}(\mathbf{W}^{[l]})}{\partial \mathbf{W}^{[l]}} \quad \text{for} \quad l = L-1, ..., 1. \tag{16}$$

The computation of the gradients in the previous layer build on the particular gradients in the post layer. Thus, some particular gradients of the expended chain rule expression have to be evaluated only once. Specifying Equations (15) and (16), by considering the quadratic cost function, the single gradients can be solved and simplified as [14]

$$\frac{\partial C(\mathbf{W}^{[L]})}{\partial \mathbf{W}^{[L]}} = \left(\underline{a}^{[L-1]}\right)^T \left((\underline{a}^{[L]} - \underline{\tilde{s}})\sigma'(z^{[L]})\right), \tag{17}$$

$$\frac{\partial C(\mathbf{W}^{[l]})}{\partial \mathbf{W}^{[l]}} = \left(\underline{a}^{[l-1]}\right)^T \left(\left((\underline{a}^{[l+1]} - \underline{\tilde{s}})\sigma'(z^{[l+1]})\mathbf{W}^{[l+1]}\right) \odot \sigma'(\underline{z}^{[l]})\right) \quad \text{for} \quad l = L-1, ..., 1, \tag{18}$$

where $\sigma'(\cdot)$ denotes the derivation of the activation function in the current layer. Equation (18) is executed iteratively for subsequent layers. The final gradients are used to update the weights by shifting the old weight values towards negative direction of the gradient, given by the Equation (19). The parameter α denotes the learning rate and is adaptively adjusted by the adaptive moment estimation (ADAM) [15] in order to enable the gradient algorithm to move gently towards the global minimum.

$$\mathbf{W}^{[l]}_{\text{new}} \leftarrow \mathbf{W}^{[l]}_{\text{old}} - \alpha \frac{\partial C}{\partial \mathbf{W}^{[l]}} \tag{19}$$

3. Experimental Verification and Discussion

This section outlines the measurement setup, as used to evaluate the nonlinear compensation performance of the following four NLE architectures on basis of identical offline data: VNLE, MP-VNLE and two types of DNN-NLEs.

3.1. Measurement Setup

A coherent single carrier transmission system over a single mode fiber (SSMF) is employed to experimentally evaluate the performance of the proposed schemes. The dual polarization 600 Gb/s experimental setup including the offline DSP stack is shown in Figure 4. The measurements were performed BtB at 1550 nm with amplified spontaneous emission (ASE) noise loading, in order to compare bit error rates before applying any forward error correction (pre-FEC BER) at varying optical signal to noise ratios (OSNR). For a net bitrate of 600 Gb/s, 704 Gb/s of pseudo random data including overheads of FEC and training have been transmitted at 88 GBd for 16QAM, at 70 GBd for 32QAM and at 68 GBd for 64QAM.

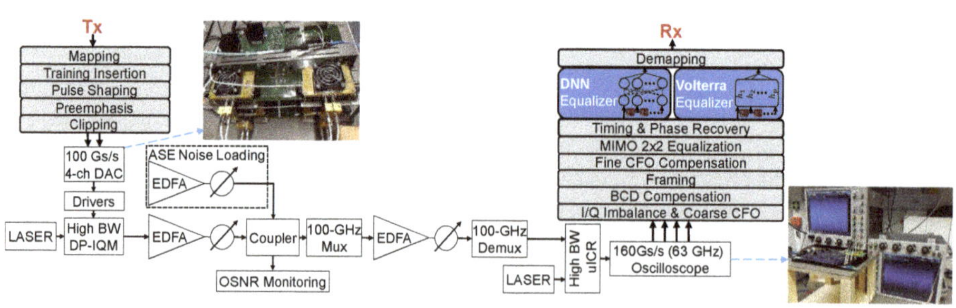

Figure 4. Back-to-back offline measurement setup including Tx and Rx digital signal processing (DSP) with different methods to compute the L-values.

The photos show a VEGA 100 GSa/s 4-channel DAC by Micram with 40 GHz bandwidth and 4.5 ENOB and two 2-channel 160 GSa/s Keysight Infiniium oscilloscopes with 63 GHz bandwidth. Their nonlinear effects are mixed with nonlinear distortions from SHF 804 A drivers, a Fujitsu 64 GBd DP-IQM and a NeoPhotonics 64 GBd class-40 HB-μICR (40 GHz). Unlike in this BtB setup, longer fiber spans as well as a WDM-setup would introduce further nonlinearities, such as single, cross and multichannel nonlinear interferences.

The transmitter DSP inserts a CAZAC (constant amplitude zero autocorrelation) based training sequence. These training symbols are used for framing, carrier frequency offset estimation, 2×2 MIMO equalization and residual chromatic dispersion compensation. These sequences of 4096 training symbols are sent prior to more than 10^5 payload data symbols repetitively. This makes it a minimal frame structure, of which the oscilloscope's memory stores up to four per capture. The DSP uses a linear preemphasis filter on the TX side against linear O/E component distortions.

The receiver DSP stack includes classical signal recovery blocks [17,18]. The different NLEs are added after the timing and phase recovery and run at one sample per symbol. For fair performance comparisons, the different NLE types operate on identical power normalized data. Regardless of the type of NLE in use, training on the particular nonlinearities is essential before deployment. CAZAC sequences do not capture nonlinearities very well. Hence, NLE training is done instead upon the payload of the first captured frame for one of the OSNR captures. Once trained, the NLE can equalize all other captures without further training.

3.2. Measurement Results over an Optical Channel with 600 Gb/s/λ

For the ultimate goal of reaching 600 Gb/s net data rate with the lowest possible BER over a wide range of OSNR levels, an optimal combination of modulation scheme and baud rate needs to be chosen in a first step, before improving its performance further with nonlinear equalization. Figure 5 shows the hard-decision preFEC BER for different modulation schemes with varying noise loading without applying NLE compensation.

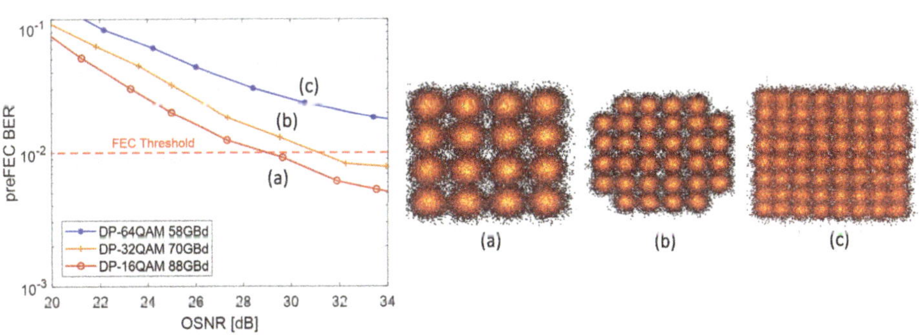

Figure 5. Back to back 600 Gb/s (+15% FEC) measurements of different modulations without nonlinear equalizers (NLE).

The horizontal red dashed line indicates a typical FEC limit of 1×10^{-2}. It can be observed that for the given setup, smaller constellation sizes improve performance much more than symbol rate reductions. DP-64QAM/58 GBd does not even reach the FEC threshold within the observed OSNR range without NLE. DP-16QAM/88 GBd proves to be the most performant scheme and thus, it is chosen for NLE architecture comparisons upon its offline data.

For fair comparisons, all options apply 4 memory taps. Per option, calibration is done exactly once on the first of four frames of the capture taken at 29.7 dB OSNR (calibration point). This ensures strict separation of calibration and measurement data, respectively training and testing data.

Figure 6 depicts the training and evaluation process. Each capture from the oscilloscope includes up to four frames, represented by the squares. For each capture the DAC is reloaded with pseudo random data. The blue square indicates the training data, while the green squares indicate the testing data. The gray colored frames are not used. For each OSNR value the overall BER is evaluated over four frames.

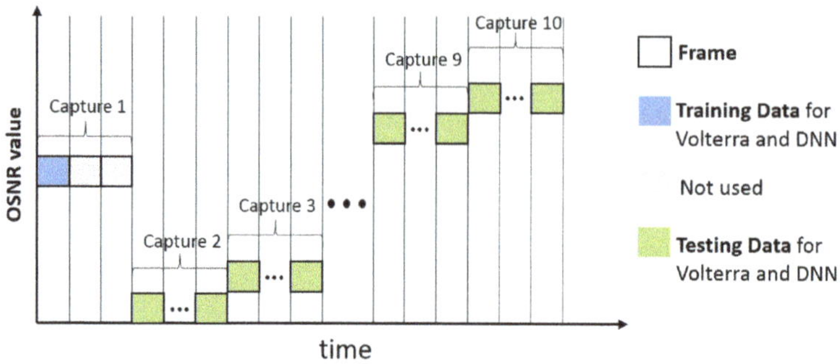

Figure 6. Overview training and evaluation process.

3.2.1. General and Memory Polynomial (MP-) Volterra nonlinear Equalizer (VNLE)

Figure 7 shows the performance improvement by applying general VNLE full kernel architectures of different orders. The gray dashed line depicts the theoretical upper limit for DP-16QAM at 88 Gbaud over an additive white Gaussian noise (AWGN) channel. The computational effort of a full kernel architecture gets extremely high with rising orders. Thus, only up to 6th order kernels are considered and evaluated. On the left hand side, Figure 7a plots the preFEC BER related to the OSNR. On the right hand side, Figure 7b shows the improvements of these base-line curves. The gains in OSNR at their crossing with the FEC threshold are plotted against the order. The general VNLE of 5th order improves the non-equalized baseline curve by 2.3 dB at a target FEC BER of 1×10^{-2}. Increasing the VNLE order further to 6th order does not yield significant additional gain.

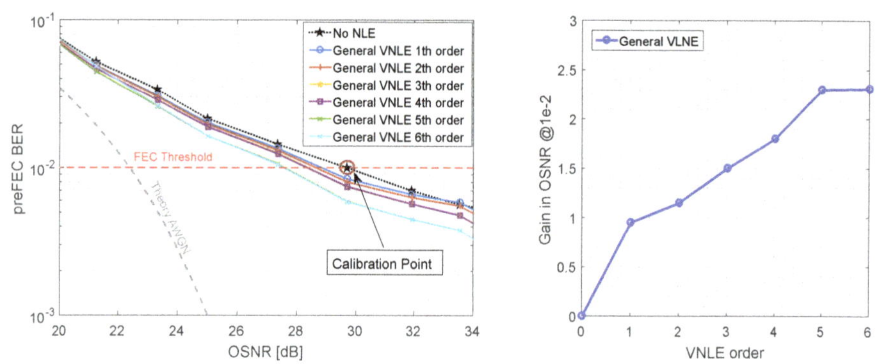

(**a**) Equalization results of general Volterra nonlinear Equalizers (VNLEs)

(**b**) VNLE gain in optical signal to noise ratio (OSNR) at FEC threshold

Figure 7. optical back-to-back 600 Gb/s/λ general VNLE preFEC measurement results.

Figure 8a shows the performance of the reduced size alternative, the MP-VNLE, which operates only on subset of kernels. As before, the performance was evaluated for different maximum polynomial orders. In comparison to the general VNLE, the architecture of the MP-VNLE is less complex and hence the computational effort to evaluate the kernels is lower. Therefore, higher orders are applicable for real systems. The MP-VNLE of 5th order improves the baseline curve by 2.0 dB at the FEC threshold of 1×10^{-2} shown in Figure 8b. It can be observed, that the reduction costs performance of 0.3 dB and that higher orders do not improve the performance. On the contrary, the performance is rather inferior. This indicates that the higher order computations lead to numerical instability when identifying and extracting the VNLE kernels with the least squares (LS) solver approach. However, besides the numerical instability such high orders are out of scope due too high implementation complexity.

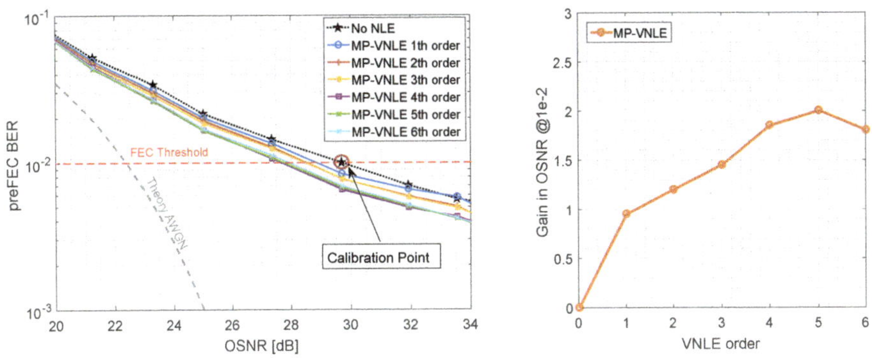

(a) Equalization results of MP-VNLEs (b) MP-VNLE gain in OSNR at FEC threshold

Figure 8. optical back-to-back 600 Gb/s/λ MP-VNLE preFEC measurement results.

3.2.2. Deep Neural Network Equalizer

While the LS-solver calibrates (MP-)VNLE kernels in a single shot, the weights and biases for DNN-NLE are trained in numerous iterations (epochs). The weights and biases of the DNN-NLE are updated by shifting the previous values towards the negative direction of the gradient. Therefore, the mini-batch backpropagation algorithm, which computes the gradients of the cost function in respect to the network parameters, and the stochastic gradient descent optimizer ADAM are applied. The hyper parameters are constant for all structures. The mini-batch size is set to 1% of the numbers of symbols of the training frame, respectively to 10^3 symbols. The learning rate of the ADAM optimizer is set to 10^{-4}, which is a common value. In order to prevent overfitting, the performance of the DNN is repetitively validated on testing data during the training phase.

The design options for DNN-NLEs are numerous and interrelated. There are no rules for the number of layers or neurons per layer to model a given problem. The more neurons the network interconnects, the finer the modeling capabilities become. More complexity, however, makes DNNs prone to overfitting [19]. Figure 9 shows two DNN equalizer architectures with multiple hidden layers and extra input nodes in order to consider the channel memory effects. On the left hand side, four separate DNNs are depicted with one output node for handling the inphase and quadrature components and polarization independently. On the right hand side, a DNN structure is shown that handles the inphase and quadrature for each polarization jointly. In principle, the joint option could learn as well nonlinear phase impairments in addition to nonlinearities. In both cases, the input nodes feed $x_I(n)$ and $x_Q(n)$ as well as $y_I(n)$ and $y_Q(n)$. Memory effects of channel and components are considered by adding time delayed versions of the input signal. The memory depth is defined by the parameter M and is set to 4, equal to the previous setting of the Volterra equalizers. The number of hidden layers layers and the corresponding number of neurons per layer of the particular networks were varied during the examination. In order to compare learning speed and performance. Table 1 list the different structures, where the last column represents the total number of independent hardware multpliers to process the data stream of two polarizations. Regardless of the structure, for the hidden neurons as well as for the output neurons a tanh has been chosen as activation functions. The design e.g., 5|20|30|40|1 stands for 5 input neurons (input signal + 4 memory taps) followed by three hidden layers with 20, 30 and 40 neurons, feeding into 1 output node. The higher the number of layers and neurons, the higher the computational effort and hence the complexity.

Table 1. DNN structures.

Number	Architecture	Design	Multipliers
1	joint I&Q	10|20|30|40|2	$(10*20+20*30+30*40+40*2)*2=4160$
2	Independent I&Q	5|20|30|40|1	$(5*20+20*30+30*40+40*1)*2*2=7760$
3	Independent I&Q	5|20|20|1	$(5*20+20*20+20*1)*2*2=2080$
4	Independent I&Q	5|10|10|10|1	$(5*10+10*10+10*10+10*1)*2*2=1040$

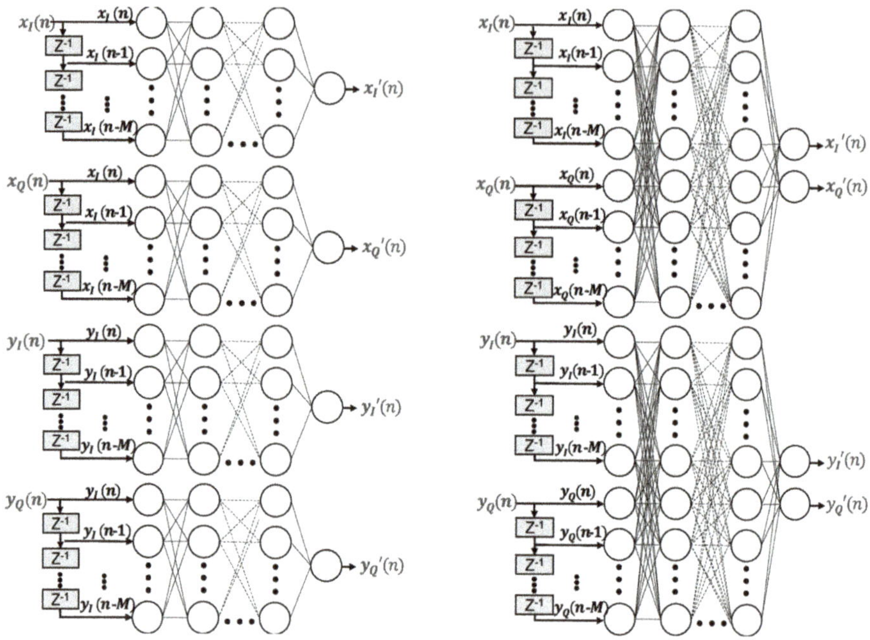

(**a**) Independent I&Q-DNN equalizer structure (**b**) joint DNNs for I&Q

Figure 9. Deep Neural Network Equalizer Architectures.

Figure 10 shows the learning process of the four different independent I&Q-DNN equalizer structures for the optical BtB 600 Gb/s measurements. The blue and the red dashed curves indicate the cost functions, given by Equation (14), of the inphase and quadrature components for one polarization, while the green curves represent the preFEC BERs related to numbers of epochs. It can be observed, that the most complex design, shown in Figure 10b converges the fastest. The learning speed of the structures depicted in Figure 10a,c,d is slightly slower, especially at the beginning. However, all options need at least around 10^5 epochs for learning the component nonlinearities. The long training process is one disadvantage of the neural network equalizer and calls for further improvement and research.

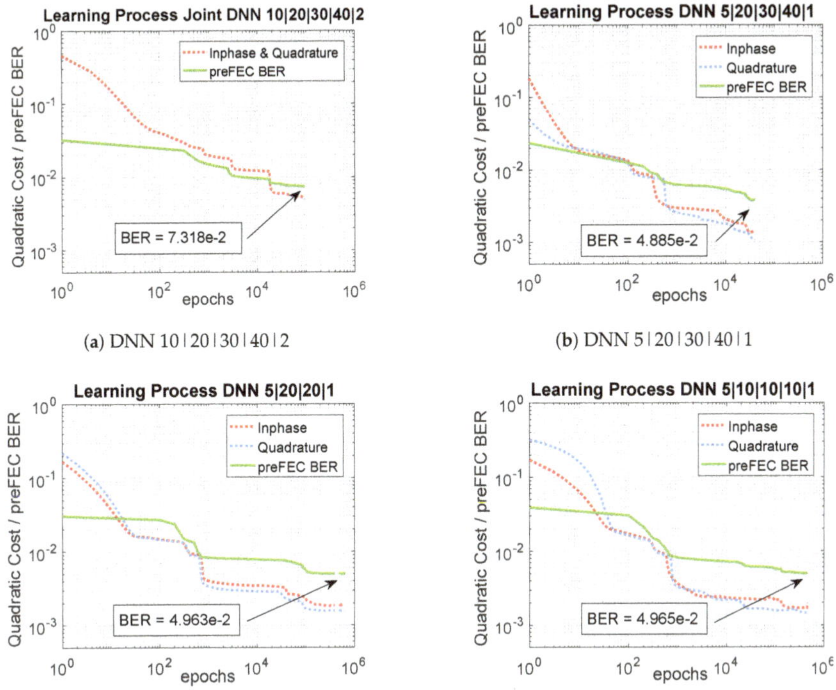

(**a**) DNN 10 | 20 | 30 | 40 | 2

(**b**) DNN 5 | 20 | 30 | 40 | 1

(**c**) DNN 5 | 20 | 20 | 1

(**d**) DNN 5 | 10 | 10 | 10 | 1

Figure 10. Learning Process of three different DNN structures.

In addition to the learning behavior, Figure 11 shows the altered constellation diagrams of the equalized signal for one polarization during the training of the DNN 5|10|10|10|1 NLE. The resulting constellations, such as the one in Figure 11f, differ fundamentally from a classical constellation. The DNN-NLE has the ability of concentrating the constellation points and to reduce the possible outputs (dimensional space) to the target 16QAM points, in order to reach a low value of the cost function. In combination with the tanh activation function, a square grid constellation occurs, which exhibits non-Gaussian distributed noise.

The overall performance of the DNN-Equalizer is shown in Figure 12. On the left hand side, Figure 12a plots the preFEC BER as a function of the OSNR, comparing the best performing general VNLE of 5th order, the best performing reduced size alternative MP-VNLE of 5th order and the proposed DNN-NLE structures. On the right hand side, Figure 12b shows the improvements of the base-line curve. The gains in OSNR at the FEC threshold are plotted. The most complex DNN 5|20|30|40|1 design, which processes I and Q separately, performs best. This architecture improves the base line curve by 2.81 dB and outperforms the 5th order VNLE and MP-VNLE by 0.53 dB and 0.80 dB, respectively. The performance losses of the lower complex 5|20|20|1 and 5|10|10|10|1 designs are very low. The performance is almost identical to the much larger design. The 10|20|30|40|1 design with joint processing, performs worst among all DNN structures. The higher number of input nodes together with the equal number of neurons in the hidden layers as the independent I and Q equalizer structure leads to suboptimal projection of the input signal.

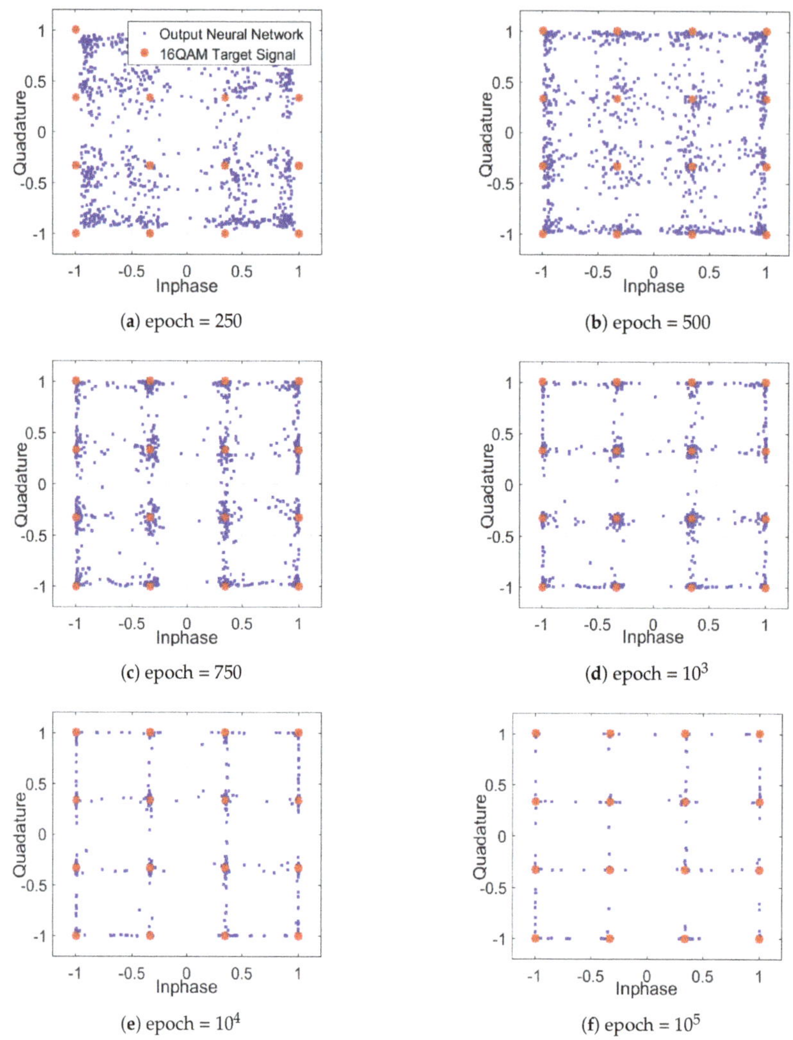

(a) epoch = 250
(b) epoch = 500
(c) epoch = 750
(d) epoch = 10^3
(e) epoch = 10^4
(f) epoch = 10^5

Figure 11. Constellation diagrams during the learning process.

Fair complexity comparisons are not trivial. Given 4 memory taps, a 5th order MP-VNLE requires 25 kernels, a general VNLE 3905 kernels [8]. However, most of these VNLE kernels can be safely pruned without severe performance reduction [20]. The 5|10|10|10|1 DNN-NLE requires 291 trainable parameters (weights, biases), respectively 1040 independent hardware multipliers, which can be pruned considerably, too [21]. Besides the sheer number of trainable parameters, the way of using them for polynomials or activation functions is of course decisive, too.

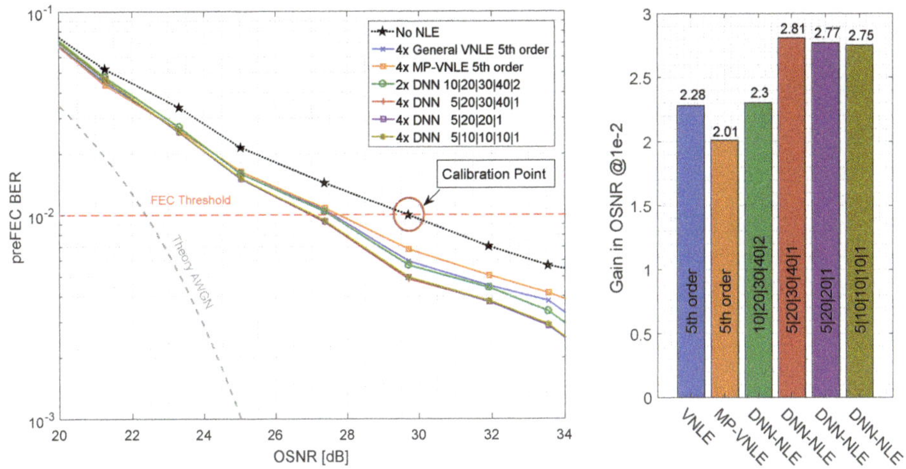

(**a**) PreFEC BER related to OSNR (**b**) Gain in OSNR at FEC threshold

Figure 12. Optical Back-to-Back 660 Gb/s/λ postFEC BER measurements results applying different Volterra and deep neural network architectures.

4. Conclusions

The algorithmic deficits in kernel identification for Volterra nonlinear equalizers and their general large architectural complexity have motivated investigations in deep neural network nonlinear equalizer alternatives. In coherent 88 Gbaud DP-16QAM 600 Gb/s measurements, deep neural network nonlinear equalizers proved to reflect systematic nonlinearities more accurately than 5th order memory polynomial and full kernel Volterra nonlinear equalizers. They outperform by 0.5 dB and 0.8 dB, respectively. Based on back-to-back measurements, where optical and electrical components nonlinearities predominate. The disadvantages of deep neural network nonlinear equalizers include their vast parameter space, which is hard to manage and their long training process. While this article highlights performance gains of DNN NLEs over classical Volterra schemes, their implementation complexity has yet to be compared in greater detail in future work, in order to demonstrate its overall superiority.

Author Contributions: Conceptualization, M.S., C.B., M.K. and S.C.; Data curation, C.B. and F.P.; Formal analysis, M.S.; Methodology, M.S.; Resources, F.P.; Software, M.S.; Supervision, M.K.; Visualization, M.S. and C.B.; Writing—original draft, M.S. and C.B.; Writing—review & editing, M.K., C.B., M.K., S.C. and S.P.

Funding: This research received no external funding.

Conflicts of Interest: The authors declare no conflict of interest.

References

1. Bluemm, C.; Schaedler, M.; Kuschnerov, M.; Pittala, F.; Xie, C. Single Carrier vs. OFDM for Coherent 600Gb/s Data Centre Interconnects with Nonlinear Equalization. In Proceedings of the 2019 Optical Fiber Communications Conference and Exhibition (OFC), San Diego, CA, USA, 3–7 March 2019.
2. Amari, A.; Dobre, O.A.; Venkatesan, R.; Kumar, O.S.; Ciblat, P.; Jaouen, Y. A survey on fiber nonlinearity compensation for 400 Gb/s and beyond optical communication systems. *IEEE Commun. Surv. Tutor.* **2017**, *19*, 3097–3113. [CrossRef]
3. Rezania, A.; Cartledge, J.C.; Bakhshali, A.; Chan, W.Y. Compensation schemes for transmitter-and receiver-based pattern-dependent distortion. *IEEE Photonics Technol. Lett.* **2016**, *28*, 2641–2644. [CrossRef]

4. Cartledge, J.C. Volterra Equalization for Nonlinearities in Optical Fiber Communications. In *Signal Processing in Photonic Communications*; Optical Society of America: Washington, DC, USA, 2017.
5. Schetzen, M. Theory of pth-order inverses of nonlinear systems. *IEEE Trans. Circuits Syst.* **1976**, *23*, 285–291. [CrossRef]
6. Kim, J.; Konstantinou, K. Digital predistortion of wideband signals based on power amplifier model with memory. *Electron. Lett.* **2001**, *37*, 1417–1418. [CrossRef]
7. Kamalov, V.; Jovanovski, L.; Vusirikala, V.; Zhang, S.; Yaman, F.; Nakamura, K.; Inoue, T.; Mateo, E.; Inada, Y. Evolution from 8QAM live traffic to PS 64-QAM with neural-network based nonlinearity compensation on 11000 km open subsea cable. In Proceedings of the Optical Fiber Communication Conference, San Diego, CA, USA, 11–15 March 2018.
8. Guan, L. *FPGA-based Digital Convolution for Wireless Applications*; Springer International Publishing: New York, NY, USA, 2017.
9. Ghannouchi, F.M.; Hammi, O.; Helaoui, M. *Behavioral Modeling and Predistortion of Wideband Wireless Transmitters*; John Wiley and Sons: Hoboken, NJ, USA, 2015.
10. Raich, R.; Zhou, G.T. Orthogonal polynomials for complex Gaussian processes. *IEEE Trans. Signal Process.* **2004**, *52*, 2788–2797. [CrossRef]
11. Musumeci, F.; Rottondi, C.; Nag, A.; Macaluso, I.; Zibar, D.; Ruffini, M.; Tornatore, M. An overview on application of machine learning techniques in optical networks. *IEEE Commun. Surv. Tutor.* **2018**, *21*, 1383–1408. [CrossRef]
12. Khan, F.N.; Fan, Q.; Lu, C.; Lau, A.P.T. An Optical Communication's Perspective on Machine Learning and Its Applications. *J. Light. Technol.* **2019**, *37*, 493–516. [CrossRef]
13. Schaedler, M.; Kuschnerov, M.; Pachnicke, S.; Bluemm, C.; Pittala, F.; Changsong, X. Subcarrier Power Loading for Coherent Optical OFDM optimized by Machine Learning. In Proceedings of the Optical Fiber Communication Conference, San Diego, CA, USA, 3–7 March 2019.
14. Bishop, C.M. *Pattern Recognition and Machine Learning*; Springer: New York, NY, USA, 2006.
15. Kingma, D.P.; Ba, J. Adam: A method for stochastic optimization. *arXiv* **2014**, arXiv:1412.6980.
16. Hecht-Nielsen, R. Theory of the backpropagation neural network. In *Neural Networks for Perception*; Academic Press: Cambridge, MA, USA, 1992; pp. 65–93.
17. Pittalà, F.; Cano, I.N.; Bluemm, C.; Schaedler, M.; Calabrò, S.; Goeger, G.; Brenot, R.; Xie, C.; Shi, C.; Liu, G.N.; et al. 400Gbit/s DP-16QAM Transmission over 40km Unamplified SSMF with Low-Cost PON Lasers. *IEEE Photonics Technol. Lett.* **2019**, *1*, 1229–1232. [CrossRef]
18. Gao, Y.; Lau, A.P.T.; Lu, C.; Wu, J.; Li, Y.; Xu, K.; Li, W.; Lin, J. Low-complexity two-stage carrier phase estimation for 16-QAM systems using QPSK partitioning and maximum likelihood detection. In Proceedings of the 2011 Optical Fiber Communication Conference and Exposition and the National Fiber Optic Engineers Conference, Los Angeles, CA, USA, 6–10 March 2011.
19. Goodfellow, I.; Bengio, Y.; Courville, A. *Deep Learning*; MIT Press: Cambridge, MA, USA, 2016; Chapter 5.
20. Huang, W.J.; Chang, W.F.; Wei, C.C.; Liu, J.J.; Chen, Y.C.; Chi, K.L.; Wang, C.L.; Shi, J.W.; Chen, J. 93% complexity reduction of Volterra nonlinear equalizer by L 1-regularization for 112-Gbps PAM-4 850-nm VCSEL optical interconnect. In Proceedings of the 2018 Optical Fiber Communications Conference and Exposition (OFC), San Diego, CA, USA, 11–15 March 2018.
21. Han, S.; Mao, H.; Dally, W.J. Deep compression: Compressing deep neural networks with pruning, trained quantization and huffman coding. *arXiv* **2015**, arXiv:1510.00149.

applied sciences

MDPI

Article

Tunable Optoelectronic Chromatic Dispersion Compensation Based on Machine Learning for Short-Reach Transmission

Stenio M. Ranzini [1,*], Francesco Da Ros [1], Henning Bülow [2] and Darko Zibar [1]

1 DTU Fotonik, Technical University of Denmark, DK-2800 Lyngby, Denmark; fdro@fotonik.dtu.dk (F.D.R.); dazi@fotonik.dtu.dk (D.Z.)

2 Nokia Bell Labs, Lorenzstr. 10, 70435 Stuttgart, Germany; henning.buelow@nokia-bell-labs.com

* Correspondence: smara@fotonik.dtu.dk

Received: 11 September 2019; Accepted: 8 October 2019; Published: 15 October 2019

Abstract: In this paper, a machine learning-based tunable optical-digital signal processor is demonstrated for a short-reach optical communication system. The effect of fiber chromatic dispersion after square-law detection is mitigated using a hybrid structure, which shares the complexity between the optical and the digital domain. The optical part mitigates the chromatic dispersion by slicing the signal into small sub-bands and delaying them accordingly, before regrouping the signal again. The optimal delay is calculated in each scenario to minimize the bit error rate. The digital part is a nonlinear equalizer based on a neural network. The results are analyzed in terms of signal-to-noise penalty at the KP4 forward error correction threshold. The penalty is calculated with respect to a back-to-back transmission without equalization. Considering 32 GBd transmission and 0 dB penalty, the proposed hybrid solution shows chromatic dispersion mitigation up to 200 ps/nm (12 km of equivalent standard single-mode fiber length) for stage 1 of the hybrid module and roughly double for the second stage. A simplified version of the optical module is demonstrated with an approximated 1.5 dB penalty compared to the complete two-stage hybrid module. Chromatic dispersion tolerance for a fixed optical structure and a simpler configuration of the nonlinear equalizer is also investigated.

Keywords: chromatic dispersion; short-reach communication; neural network; hybrid signal processing

1. Introduction

The well-known increase in information rate is a particular concern for inter-data center communication due to chromatic dispersion (CD). Increasing the rate of information reduces the optimal reachable transmission [1]. Using an O-band for standard single-mode fibers, where the CD can almost be neglected, is one way to avoid needing to mitigate it. Nonetheless, transmission in the O-band has high attenuation, which reduces the length of the transmission link [2]. Alternatively, solutions in the C-band, which address the intersymbol interference induced by dispersion, can also be considered [3–5].

Coherent and direct detection (DD) systems are both possible technologies for C-band short-reach communication [3,6]. A coherent receiver has access to the amplitude and phase of the signal. Adding it together with digital signal processing (DSP), it brings with it huge potential to mitigate transmission impairments. However, inter-data center applications require cheaper and low-power transponders, which can be a challenge in these systems.

DD-receivers are the current alternative, but CD cannot be fully compensated in the digital domain as only the power is detected in the receiver. Different options for inter- and intra-data center interconnection, as well as for mobile fronthaul application for operation on the C-band have already been discussed in recent years [4,5]. A few of them operate on PAM-4 modulation and different options for DSP-based receiver equalization, such as Volterra and maximum likelihood sequence estimation (MLSE) [7–9] and DSP-based pre-distortion, such as Tomlinson Harashima precoding and the kramers-kronig receiver [10,11]. Works with an on-off keying (OOK) modulation format have also been presented in the literature [12]. Recently, further signal processing technologies for this application range have been proposed, such as DSP-implemented neural networks [13,14] or silicon photonic-based opto-electronic reservoir computers with analog electronic processing [15].

Alternatively, optical dispersion compensation modules as fiber Bragg grating and dispersion compensation fibers are more commonly used to compensate for the CD in the optical domain. However, they are not easy to tune and need to be designed for a specific link. The latter also has issues with high attenuation and a large footprint.

Here, we propose a tuneable optoelectronic solution to mitigate the CD in DD systems. The processing structure we are investigating is located close to the approach using a simplified optical reservoir processor, discussed in [15], together with an NN digital processor [14]. This work is an extension of our preliminary results of [16]. Here, we improve the digital equalization by using a neural network equalizer and propose a simplification of the optical structure to reduce footprint. The optical module slices the signal into narrow frequency sub-bands and delays them accordingly to the CD, before recombining everything back into a single signal. Increasing the number of sub-bands improves performance, but increases complexity. Alternatively, we show that the performance can be improved using the optical component with fewer sub-bands if a further stage of equalization in the digital domain is also applied. Furthermore, a complexity reduction is demonstrated by using half of the proposed optical module structure together with an NN equalizer. This approach slightly reduces the gain, but can be attractive for inter-data center communication due to the reduced footprint. Finally, the number of neurons required for the NN equalizer is analyzed.

The paper is organized as follows. Section 2 presents the system description for short-reach communication and simulation setup. Sections 2.1 and 2.2 describe the hybrid signal process module. The former is the optical structure used to mitigate CD, and the latter describes the NN equalization and training process. Section 3 shows the simulation results starting with the complete hybrid structure and ending with a simplified version of it. Finally, in Section 4 we summarize the main conclusions of the paper.

2. System Description and Simulation Setup

We propose the following system for short-reach communication. At the transmitter, a digital-to-analog converter (DAC) maps the signal to an OOK modulation format with root-raised cosine (RRC) as pulse shaping. We believe that our method can be easily extended to a PAM-4 modulation format, which will be considered in our future works. The Mach–Zehnder modulator (MZM) is used to modulate the signal. A standard single-mode fiber (SSMF) propagates the signal, and an erbium-doped fiber amplifier (EDFA) amplifies it. At the receiver, an optical band-pass filters the noise outside the signal bandwidth. Before detection, an optical module is used to pre-distort the signal to mitigate the CD. This process will be used to assist the nonlinear digital equalizer. A photodetector (PD) is then used to detect the signal, and an analog-to-digital converter (ADC) converts it to the digital domain. A DSP with NN equalizer and RRC filter are used to equalize the signal. The optical module, PD, and NN equalizer are the proposed hybrid optoelectronic blocks used to mitigate CD for direct detection systems.

Figure 1 shows the simulation setup used to validate the proposed system. At the transmitter, the DAC is the assumed ideal and an OOK 32-GBd signal is generated with 2^{18} symbols, up-sampled to eight samples per symbol and filtered by a RRC filter (roll-off = 0.1). The digital signal is directly mapped onto the optical domain by assuming an ideal linear transformation. The MZM is assumed ideal in order to neglect its nonlinear transfer function as a source of degradation and to focus solely on the impact of CD. The CD (D = 16.4 ps/nm/km) is the only impairment considered in the optical fiber, and additive white Gaussian noise (AWGN) is used to simulate the pre-amplifier at the Rx point. The noise variance and fiber length (accumulated CD) are swept to meet the target SNR and distance. The signal is then detected by a DD receiver.

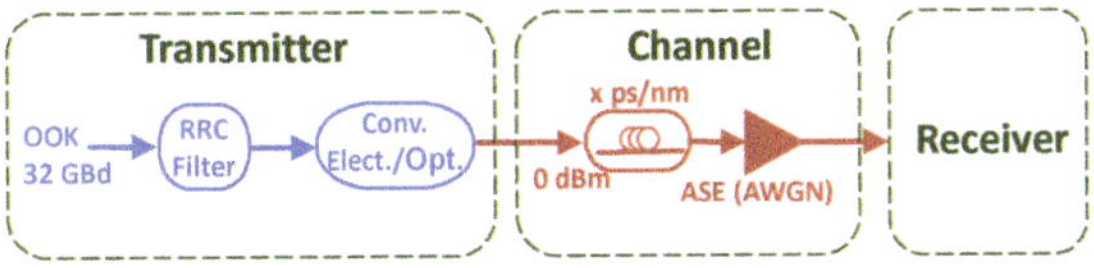

Figure 1. Simulation setup.

Figure 2 shows different topologies of the receiver that is analyzed in the result section. For all of them, a second-order Gaussian filter (40-GHz bandwidth) is used as a band-pass filter, which reduces the noise-to-noise beating in the PD. Moreover, the RRC filter block (ADC is assumed ideal) is used as a low-pass filter to downsample the signal to one sample per symbol. Receiver (a) is used as a reference. No optical pre-processing or digital equalization is applied. Receivers (b) and (c) show the block diagram for digital equalization with NN and MLSE (more details in Section 2.2), respectively. Receiver (d) shows the block diagram for optical pre-processing using a complete optical module (more details in Section 2.1). Receivers (e) and (f) show the proposed hybrid solution to mitigate CD. The former is based on the complete optical module, together with the NN. The latter is the receiver with the proposed simplified optical module together with an NN, respectively.

The simulation results firstly show a comparison between the performance of optical equalization and electronic equalization, individually. This results are followed by the complete hybrid module. Secondly, we analyze the requirements of the optical module to mitigate CD in a hybrid system. Thirdly, we compare a simplified hybrid module with the complete hybrid module. Fourthly, we study the tolerance of the simplified and complete hybrid module to CD. And lastly, we investigate the performance of NN in both hybrid systems.

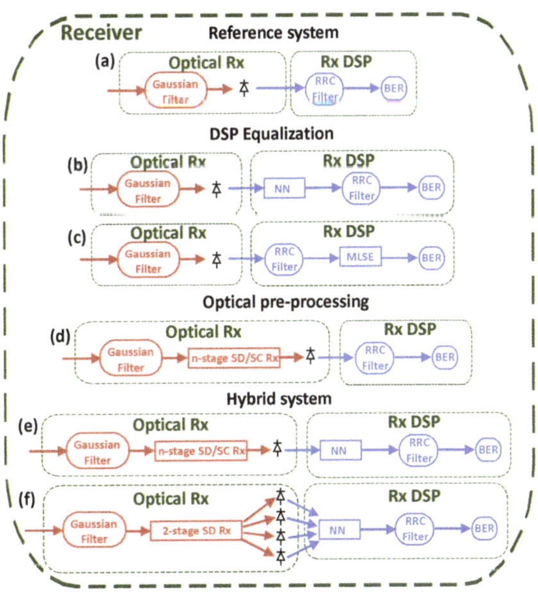

Figure 2. Receiver configurations.

2.1. Optical Pre-Processing

The optical module to mitigate the CD is based on the spectrum decomposition (SD) and spectral composition (SC) blocks [17]. SD is a technique in which the signal is sliced into narrow frequency sub-bands. The basic building-block for this process is a phase-tunable Mach–Zehnder delay interferometer (MZDI). Figure 3a is an example of a two-stage SD (four sub-bands). The orange dashed lines highlight the MZDI. The bandwidth of the sub-bands and the relative position of each other depends on the delay (Δt_i) and phase shift ($\Delta\phi_i$) of the MZDI. i is the number of stages. Equations (1) and (2) describe it.

$$\Delta t_i = \frac{1}{2 \cdot d_i} \quad (1)$$

$$\Delta\phi_i = \pi + k2\pi + \omega_i \Delta t_i,\ k \in \mathbb{Z} \quad (2)$$

in which ω is the angular frequency where the signal's power is maximized in the output upper arm of the MZDI and minimized for the other. The inverse delay, d, determines the distance in frequency where the maximum and minimum of the signal's power is alternated in each output arm of the MZDI. For example, considering $\omega_1 = 2 \cdot \pi \cdot 12 \cdot 10^9$ rad/s and $d_1 = 8$ GHz, the signal's power is maximized at 12 GHz and alternates between minimum and maximum power every 8 GHz for the output upper arm, and the opposite for the lower arm. For a two-stage SD, we considered in the first stage $d_1 = 8$ GHz and $\omega_1 = 2 \cdot \pi \cdot 12 \cdot 10^9$ rad/s. For the second stage we considered, $d_2 = 16$ GHz and $\omega_2 = 2 \cdot \pi \cdot 12 \cdot 10^9$ rad/s.

SC applies the inverse transfer function of the SD, and it is highlighted in blue dashed lines in Figure 3b. Perfect reconstruction of the signal is obtained if the SD is applied, followed directly by the SC. By knowing the transfer function of CD, a time delay and phase shift in each of the sub-bands of the SD can approach the inverse of the CD transfer function. The approximation increases in accuracy for increasingly narrower spectral slices. The values of the delay and phase depend on the number of sub-bands and the CD. The SD and SC structures, together with the delay and phase to mitigate the CD in each sub-band, is called the n-stage SD/SC Rx (Figure 4a) in this work.

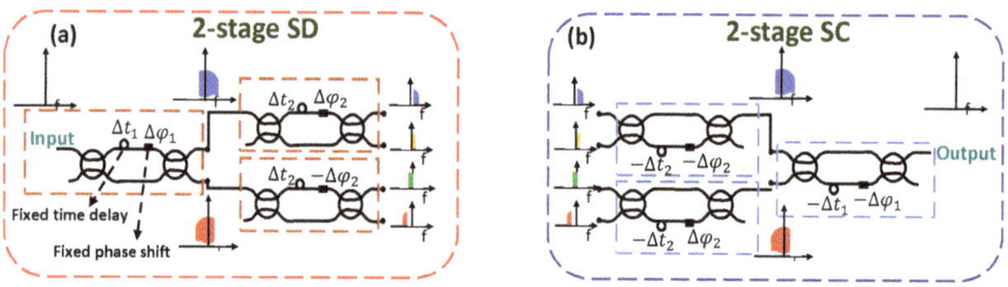

Figure 3. Optical structure to slice the signal into frequency sub-bands (**a**) and reconstruct it (**b**).

Figure 4 shows the optical pre-processing modules. Figure 4a shows an n-stage SD/SC Rx, which is composed of $2 \times (2^n - 1)$ MZDI modules and creates 2^n sub-bands. Figure 4b shows a simplified version of the optical module. We are proposing to remove the SC part in the two-stage SD/SC structure. In this configuration, the optical module is acting as an optical pre-processing structure to assist the equalization process in the digital domain. The phase shift is not considered because the signal is directly connected to the PD. The different signals in the electrical domain need to be recombined and the equalizer can perform this operation implicitly. An addition of three PDs is necessary for receiving all four optical outputs of a two-stage SD. The performance of both optical modules is analyzed in the results section.

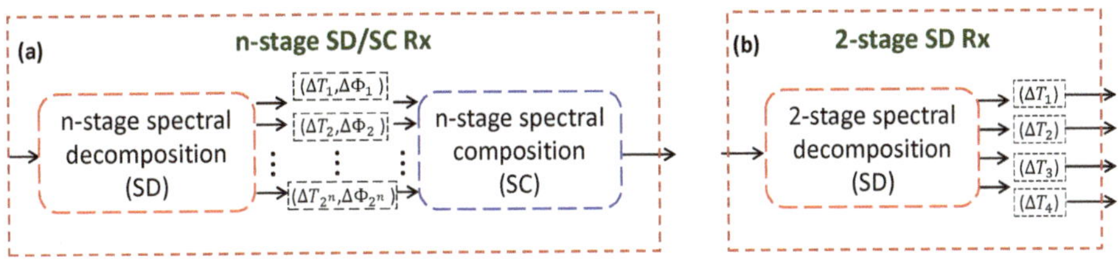

Figure 4. Optical pre-processing. (**a**) Complete optical module, (**b**) reduced optical module. Time delay and phase shift were trained to find the optimal point to mitigate CD.

2.2. Digital Processing

Figure 5a shows a schematic of an NN equalizer [18]. It is a two-layer NN, performing regression, with L neurons in the hidden layer and eight neurons in the output layer (one symbol). A hyperbolic tangent function is used as an activation function in the hidden layer, while a linear function is used in the output layer. The number of inputs of the NN depends on the optical module. If the n-stage SD/SC Rx is being analyzed, there are five symbols (40 samples) in the input. Otherwise, the number of input symbols are five per PD, which means 200 samples for the two-stage SD Rx. It is important to highlight that the middle symbol of the input is the one being equalized (symbol #3). In this way, the NN can take into account the symbols' neighbors' interference. The input window is shifted by a symbol period. The n-stage SD/SC Rx (Figure 4a), together with the NN, is called in this work the hybrid n-stage SD/SC. The two-stage SD RX (Figure 4b), together with the NN, is the hybrid two-stage SD.

The weights of NN, group delay, and phase of CD mitigation of the optical modules are jointly trained when these variables are available in the simulation. The algorithm to update them is the stochastic gradient descent with adaptive moment estimation [19]. The loss function considered is the cross-entropy between the transmitted and received bits. During an entire simulation (2^{18} symbols), all the trainable variables are kept constant and they are only updated when the next simulation starts. All the variables are updated at the same time after 600 iterations of the training process.

Figure 5b shows an example of a decided 8-bit sequence with seven memory MLSEs. The MLSE "m" memory is an algorithm with 2^m trellis states. The probability density function of the received sequence bits are first estimated, and then used to make the decision [9].

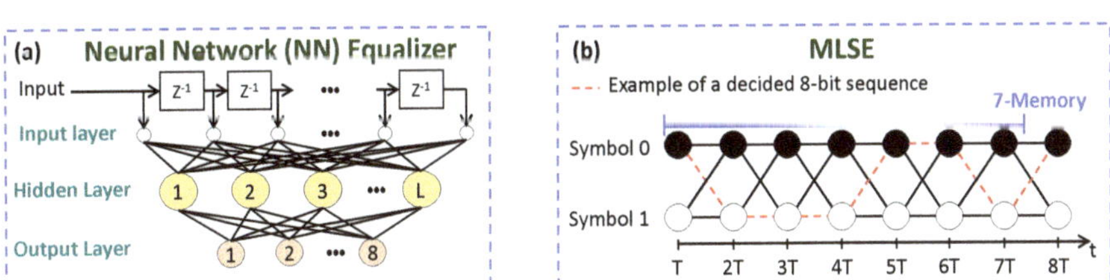

Figure 5. Digital processing algorithms used to process the signal. (**a**) NN equalizer. (**b**) MLSE.

3. Results

All the results are shown in terms of SNR penalty at KP4 FEC, with a hard-decision BER threshold of 2.26×10^{-4} [20]. The SNR varies from 13 to 20 dB. Each simulation is repeated five times to measure the statistical relevance of the results.

3.1. Hybrid Signal Processing

Figure 6 shows the results comparing the performance of a complete hybrid optoelectronic structure. Figure 6a compares the optical and electronic equalization, individually. The black curve (right-side triangle) shows the result without optical or digital signal processing, namely our reference system. The blue (circle), purple (square), and dark blue (upside-down triangle) curves are the ones with only optical pre-processing. We note that increasing the number of stages in the n-stage SD/SC structure increases the maximum transmission reach. As we better compensate the CD, the narrow slices are required [21]. Using four-stage SD/SC, a transmission of more than 574 ps/nm of accumulated CD (35 km of equivalent fiber) is demonstrated with less than 1 dB of SNR penalty at KP4, compared to a back-to-back transmission. The dark green (star) dashed line curve and yellow (upper triangle) dashed line curve are the digital processing for MLSE and NN, respectively. It is worth to point out that the seven-memory MLSE and the NN had similar performance. Using only electronic processing with an NN equalizer or MLSE, we showed a transmission of 164 ps/nm of accumulated CD (10 km of equivalent SSMF), considering a 0 dB SNR penalty at KP4 FEC compared to a back-to-back transmission. A difference between optical and digital performance can be clearly seen. As the phase is not accessible in the digital domain after the PD, such a difference appears [8].

Figure 6b compares the complete hybrid optoelectronic system with optical equalization only. The red (asterisk) and grey (diamonds) show the result for the hybrid processing. We note that it is possible to further mitigate the CD and increase the transmission reach by adding post-processing in the digital signal part. This post-equalization is mitigating the residual CD of the system. In other words, instead of adding more stages in the SD/SC structure, which increases the optical complexity, it is possible to transfer part of this complexity to the digital domain. Figure 6b shows an equivalent SSMF fiber transmission of 12 km (≈200 ps/nm) for a hybrid one-stage SD/SC module and 25 km (≈400 ps/nm) for a hybrid two-stage SD/SC module, considering a 0 dB penalty. Another observation is that the minimum SNR penalty is 0 dB for optical pre-processing, but less than 0 dB for electronic processing. This indicates that the optical module is trying to compensate only for the CD, while the electronic one is also trying to mitigate some of the effects of the square-law reception.

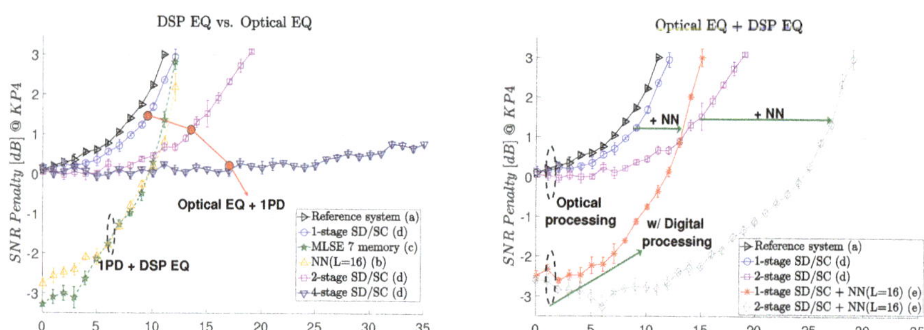

(**a**) Comparison between optical pre-processing and electronic equalization.

(**b**) Comparison between the complete hybrid system and optical pre-processing.

Figure 6. Compared performance of the complete hybrid optoelectronic system. The receiver topology considered in Figure 2 is highlighted inside the parentheses in the legend of the figures.

3.2. Delay and Phase Shift Impact on CD for Hybrid Two-Stage SD/SC

The CD transfer function shows that the modulated signal has a time delay and a phase shift effect during propagation. In this section, we investigate these effects considering the hybrid two-stage SD/SC structure.

Figure 7 shows these results. The blue curve (circle) shows the results achieved by applying only the time delay. The grey curve (diamond) shows the result of applying time delay and phase shift. These two curves show a similar performance, suggesting that the phase shift is not necessary. By applying only the phase shift (light green curve-square), instead, the performance worsens compared to using only digital equalization (yellow curve-upper triangle). We can, therefore, conclude that delay is the most significant operation in the hybrid module to mitigate CD.

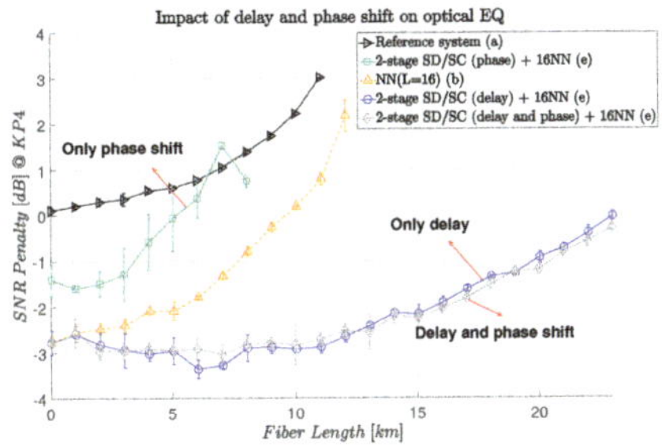

Figure 7. Simulation results to analyze the impact of delay and phase shift in the hybrid two-stage SD/SC structure. The receiver topology considered in Figure 2 is highlighted inside the parentheses in the legend of the figure.

3.3. Simplified Hybrid Processing Performance

In this section, we investigate the effects of simplifying the optical module by implicitly transferring the SC operation to the nonlinear equalizer.

Figure 8 shows a comparison between the two-stage SD/SC and two-stage SD optical modules in a hybrid topology with an NN (L = 16). Without the SC structure, a two-stage SD module still mitigates the CD with ≈1.5 dB of the SNR penalty at KP4 FEC. We can also note that in the back-to-back transmission, there is almost no SNR penalty, which indicates that the NN is being able to regroup the four signals back into a single one.

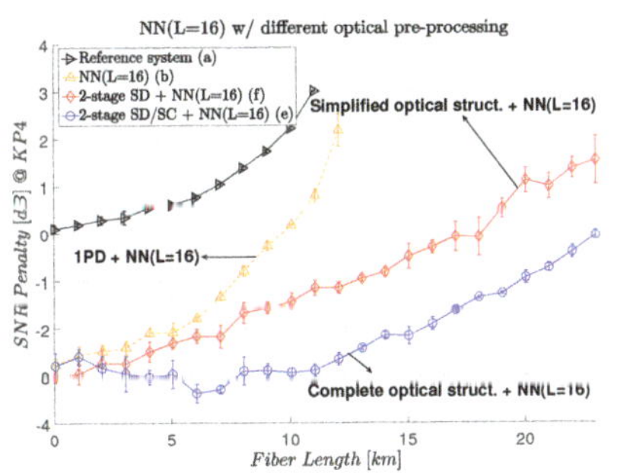

Figure 8. Simulation results to compare the hybrid two-stage SD/SC and hybrid two-stage SD/SC modules. The receiver topology considered in Figure 2 is highlighted inside the parentheses in the legend of the figure.

3.4. Delay Tolerance for Optical Hybrid Processing

One of the challenges of the proposed optical modules was finding the optimal delay to mitigate the CD. This happened because we used the stochastic gradient descent with adaptive moment estimation to update the delay and the weights of the NN at the same time. This can make the convergence process more difficult. To avoid this slow convergence and simultaneously simplify the optical structure, we investigated the performance of the hybrid systems for a fixed delay at 12 km of fiber transmission. Figure 9 shows the tolerance to CD in this scenario. The hybrid two-stage SD has a CD tolerance similar to the case in which the optimal delay is calculated. In other words, almost no penalty is introduced by fixing the delay. In contrast, the hybrid two-stage SD/SC with fixed delay showed lower CD tolerance. This might be an indication that the sensibility in the delay of the full optical structure (SC/SD) is mainly related to the SC section.

Comparing both modules with a fixed delay and 0 dB SNR penalty at KP4 FEC, the hybrid two-stage SD/SC shows a transmission of ≈370 ps/nm of accumulated CD (23 km of equivalent SSMF). The hybrid two-stage SD shows a transmission of ≈300 ps/nm of accumulated CD (18 km of equivalent SSMF).

Considering only the hybrid two-stage SD, there was no penalty for the distance from 7 to 18 km. Below 7 km, the performance with a fixed delay was higher, which could indicate a missing interpolation operation from the NN to adapt to a better time delay. For distances from 18 to 23 km, we note that the fixed delay had less penalty. However, the curve with optimal delay showed a higher standard deviation, which indicates a more difficult convergence for the trainable variables. Fixing the delay would mean simplifying the convergence for the NN and, for this specific amount of CD, it showed better performance. Considering the hybrid two-stage SD/SC, we note that there was no penalty from 8 to 13 km. Outside this range, the penalty increased faster compared to the optimal delay.

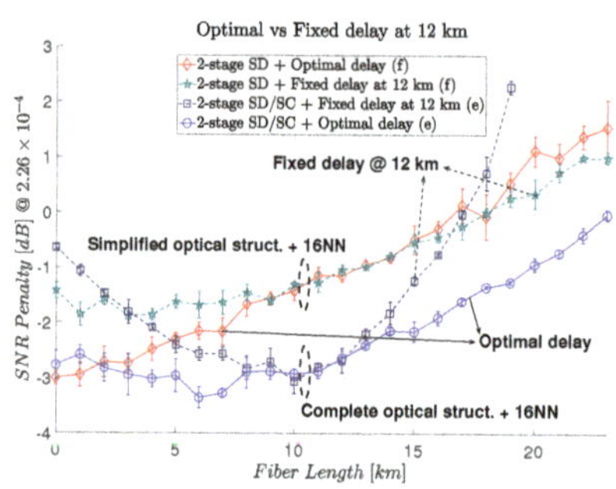

Figure 9. Simulation results when comparing optimal delay and fixed delay for 12 km of fiber length for hybrid processing. The receiver topology considered in Figure 2 is highlighted inside the parentheses in the legend of the figure.

3.5. Impact of the NN in Hybrid Processing

Figure 10 shows the impact of decreasing or increasing the number of neurons in the hidden layer for the NN. Figure 10a,b show the results for the hybrid two-stage SD/SC and hybrid two-stage SD, respectively. The delay is fixed considering a 12 km transmission. As a reference, a hybrid two-stage SD/SC with 16 neurons in the hidden layer with optimal delay is shown.

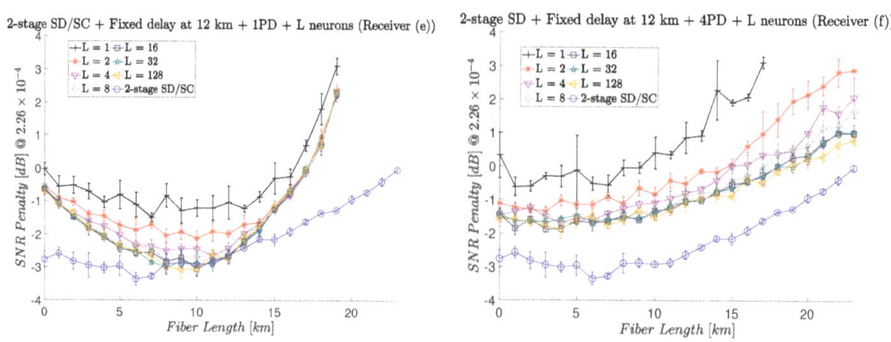

(a) two-stage SD/SC optical structure.

(b) two-stage SD optical structure.

Figure 10. Analyzing number of neurons for the digital equalization in the hybrid structure.

Figure 10a shows that between 8 and 16 neurons is the optimal number of neurons in the hidden layer. For less than 8 neurons, the SNR penalty increases. For more than 16 we cannot see significant improvements. Although the NN could potentially find and adjust the optimal required delay, this is not happening even by increasing the number of neurons in the hidden layer. Increasing the number of neurons makes the convergence process more difficult. Figure 10b also shows that the ideal number of neurons in the hidden layer is also between 8 and 16. Similar behavior as Figure 10a for using less or more neurons is repeated here. Even considering the SNR penalty, using fewer neurons is still a possibility for the inter-data center scenario, because of the reduced complexity of the nonlinear equalizer.

4. Conclusions

In this paper, we showed a tuneable hybrid signal processing system to increase the maximum transmission reach for a DD system. Considering only the optical part of the hybrid module (n-stage SD/SC), we demonstrated that adjusting only the delay in the middle of the optical structure was sufficient to mitigate the CD. The more transmission reach was desired, the more optical complex structure is required. Using a four-stage SD/SC, a transmission of more than 574 ps/nm of accumulated CD (35 km of equivalent SSMF) was demonstrated with less than 1 dB of SNR penalty at KP4, compared to a back-to-back transmission.

Using only electronic processing with an NN equalizer or MLSE, we showed a transmission with 164 ps/nm (10 km of equivalent SSMF) of accumulated CD, in the same scenario. Adding the NN equalizer as post-processing in the hybrid module, we transferred part of the complexity from the optical to the electronic domain. For a one-stage hybrid module (one-stage SD/SC + NN (L = 16)), we showed a CD mitigation of $\approx$200 ps/nm (12 km of equivalent standard single-mode fiber length). The tolerance was roughly doubled for a two-stage hybrid module (two-stage SD/SC + NN (L = 16)). Removing the SC from the hybrid structure (two-stage SD + NN (L = 16)) reduced the optical complexity module without losing the capacity of mitigating the CD. However, there was a $\approx$1.5 dB penalty compared to a complete two-stage hybrid module.

A key component to mitigate the CD in the optical modules is the time delay. We showed that it is possible to keep this delay constant for a specific range of transmission reach and simplify the optical structure. For the two-stage hybrid module and considering 0 dB SNR penalty at KP4 FEC compared to a back-to-back transmission, we showed a transmission with an accumulated CD of $\approx$370 ps/nm (23 km of equivalent SSMF) using the time delay for a 12 km transmission. For the hybrid two-stage SD, the proposal module, there was no penalty by fixing the time delay. At the same scenario, we showed a transmission with accumulated CD of $\approx$300 ps/nm (18 km of equivalent SSMF). Finally, we analyzed the required number of neurons in the hidden layer of the NN. For both structures, the ideal number calculated was between 8 and 16 neurons.

Author Contributions: Conceptualization, S.M.R., F.D.R., H.B. and D.Z.; Investigation, S.M.R.; Supervision, F.D.R., H.B. and D.Z.; Validation, S.M.R.

Funding: This project has received funding from the European Union's Horizon 2020 research and innovation programme under the Marie Skłodowska-Curie grant agreement No 766115. It is also supported by the European Research Council through the ERC-CoG FRECOM project (grant agreement no. 771878).

Conflicts of Interest: The authors declare no conflict of interest.

References

1. Agrawal, G.P. *Fiber-Optic Communication Systems*, 3rd ed.; John Wiley & Sons: Hoboken, NJ, USA, 2002.
2. Turkiewicz, J.P. Cost-effective n× 25 Gbit/s DWDM transmission in the 1310 nm wavelength domain. *Opt. Fiber Technol.* **2011**, *17*, 179–184. [CrossRef]
3. Cheng, J.; Xie, C.; Chen, Y.; Chen, X.; Tang, M.; Fu, S. Comparison of Coherent and IMDD Transceivers for Intra Datacenter Optical Interconnects. In Proceedings of the Optical Fiber Communications Conference and Exhibition (OFC), San Diego, CA, USA, 3–7 March 2019.

4. Chagnon, M. Direct-detection Technologies for Intra-and Inter-Data Center Optical Links. In Proceedings of the Optical Fiber Communications Conference and Exhibition (OFC), San Diego, CA, USA, 3–7 March 2019; pp. 1–41.
5. Zhou, H.; Li, Y.; Liu, Y.; Yue, L.; Gao, C.; Li, W.; Qiu, J.; Guo, H.; Hong, X.; Zuo, Y.; et al. Recent Advances in Equalization Technologies for Short-Reach Optical Links Based on PAM4 Modulation: A Review. *Appl. Sci.* **2019**, *9*, 2342. [CrossRef]
6. Zhong, K.; Zhou, X.; Huo, J.; Yu, C.; Lu, C.; Lau, A.P.T. Digital signal processing for short-reach optical communications: A review of current technologies and future trends. *J. Lightwave Technol. (JLT)* **2018**, *36*, 377–400. [CrossRef]
7. Jignesh, J.; Eriksson, T.A.; Chagnon, M.; Corcoran, B.; Lowery, A.J.; Buchali, F.; Bülow, H. Transmitter-side Volterra filtering for increased dispersion tolerance in 56 Gbaud PAM-4 systems. In Proceedings of the Optical Fiber Communications Conference and Exhibition (OFC), San Diego, CA, USA, 11–15 March 2018.
8. Rubsamen, M.; Gene, J.M.; Winzer, P.J.; Essiambre, R.J. ISI mitigation capability of MLSE direct-detection receivers. *IEEE Photonics Technol. Lett. (PTL)* **2008**, *20*, 656–658. [CrossRef]
9. Agazzi, O.E.; Hueda, M.R.; Carrer, H.S.; Crivelli, D.E. Maximum-likelihood sequence estimation in dispersive optical channels. *J. Lightwave Technol. (JLT)* **2005**, *23*, 749–763. [CrossRef]
10. Qian, H.; Mathieu, C.; Karsten, S.; Fred, B.; Henning, B. High Data Rate Tomlinson-Harashima precoding-Based Pam Transmission. In Proceedings of the European Conference on Optical Communication (ECOC), Dublin, Germany, 22–26 September 2019.
11. Mecozzi, A.; Antonelli, C.; Shtaif, M. Kramers–Kronig coherent receiver. *Optica* **2016**, *3*, 1220–1227. [CrossRef]
12. Mardoyan, H.; Jorge, F.; Baeuerle, B.; Estaran, J.M.; Heni, W.; Konczykowska, A.; Riet, M.; Duval, B.; Nodjiadjim, V.; Goix, M.; et al. 222-Gbaud On-Off Keying Transmitter Using Ultra-High-Speed 2:1-Selector And Plasmonic Modulator On Silicon Photonic. In Proceedings of the European Conference on Optical Communication (ECOC), Dublin, Germany, 22–26 September 2019.
13. Gaiarin, S.; Pang, X.; Ozolins, O.; Jones, R.T.; da Silva, E.P.; Schatz, R.; Westergren, U.; Popov, S.; Jacobsen, G.; Zibar, D. High speed PAM-8 optical interconnects with digital equalization based on neural network. In Proceedings of the Asia Communications and Photonics Conference (ACP), Wuhan, China, 2–5 November 2016.
14. Chagnon, M.; Karanov, B.; Schmalen, L. Experimental demonstration of a dispersion tolerant end-to-end deep learning-based im-dd transmission system. In Proceedings of the European Conference on Optical Communication (ECOC), Roma, Italy, 23–27 September 2018; pp. 1–3.
15. Katumba, A.; Yin, X.; Dambre, J.; Bienstman, P. A Neuromorphic Silicon Photonics Nonlinear Equalizer For Optical Communications With Intensity Modulation and Direct Detection. *J. Lightwave Technol. (JLT)* **2019**, *37*, 2232–2239. [CrossRef]
16. Ranzini, S.M.; da Ros, F.; Zibar, D. Joint low-complexity opto-electronic chromatic dispersion compensation for short-reach transmission. In Proceedings of the IEEE Photonics Conference (IPC), San Antonio, TX, USA, 30 September–3 October 2019; pp. 1–2.
17. Hillerkuss, D.; Schmogrow, R.; Schellinger, T.; Jordan, M.; Winter, M.; Huber, G.; Vallaitis, T.; Bonk, R.; Kleinow, P.; Frey, F.; et al. 26 Tbit s- 1 line-rate super-channel transmission utilizing all-optical fast Fourier transform processing. *Nat. Photonics* **2011**, *5*, 364. [CrossRef]
18. Chen, S.; Gibson, G.; Cowan, C.; Grant, P. Adaptive equalization of finite non-linear channels using multilayer perceptrons. *Signal Process.* **1990**, *20*, 107–119. [CrossRef]
19. Kingma, D.P.; Ba, J. Adam: A method for stochastic optimization. *arXiv* **2014**, arXiv:1412.6980.
20. Agrell, E.; Secondini, M. Information-theoretic tools for optical communications engineers. In Proceedings of the IEEE Photonics Conference (IPC), Reston, VA, USA, 30 September–4 October 2018; pp. 1–5.
21. Slim, I.; Mezghani, A.; Baltar, L.G.; Qi, J.; Hauske, F.N.; Nossek, J.A. Delayed single-tap frequency-domain chromatic-dispersion compensation. *IEEE Photonics Technol. Lett. (PTL)* **2012**, *25*, 167–170. [CrossRef]

applied sciences

MDPI

Article

Optimization Algorithms of Neural Networks for Traditional Time-Domain Equalizer in Optical Communications

Haide Wang [1], Ji Zhou [1,*], Yizhao Wang [1], Jinlong Wei [2], Weiping Liu [1], Changyuan Yu [3] and Zhaohui Li [4,5,*]

1 Department of Electronic Engineering, College of Information Science and Technology, Jinan University, Guangzhou 510632, China; 1834041007@stu2018.jnu.edu.cn (H.W.); wyz0714@stu2017.jnu.edu.cn (Y.W.); wpl@jnu.edu.cn (W.L.)

2 Huawei Technologies Duesseldorf GmbH, European Research Center, 80992 Munich, Germany; jinlongwei@gmail.com

3 Department of Electronic and Information Engineering, The Hong Kong Polytechnic University, Hong Kong, China; changyuan.yu@polyu.edu.hk

4 State Key Laboratory of Optoelectronic Materials and Technologies, School of Electronics and Information Technology, Sun Yat-sen University, Guangzhou 510275, China

5 Southern Marine Science and Engineering Guangdong Laboratory (Zhuhai), Zhuhai 519082, China

* Correspondence: zhouji@jnu.edu.cn (J.Z.); lzhh88@sysu.edu.cn (Z.L.)

Received: 20 August 2019; Accepted: 15 September 2019; Published: 18 September 2019

Abstract: Neural networks (NNs) have been successfully applied to channel equalization for optical communications. In optical fiber communications, the linear equalizer and the nonlinear equalizer with traditional structures might be more appropriate than NNs for performing real-time digital signal processing, owing to its much lower computational complexity. However, the optimization algorithms of NNs are useful in many optimization problems. In this paper, we propose and evaluate the tap estimation schemes for the equalizer with traditional structures in optical fiber communications using the optimization algorithms commonly used in the NNs. The experimental results show that adaptive moment estimation algorithm and batch gradient descent method perform well in the tap estimation of equalizer. In conclusion, the optimization algorithms of NNs are useful in the tap estimation of equalizer with traditional structures in optical communications.

Keywords: neural networks; optical communications; optimization; equalizer; tap estimation

1. Introduction

In recent years, since deep learning has been applied in image recognition, natural language processing, target tracking, recommendation system, and so on, it is becoming one of the most popular spots for academic research and industrial application [1]. Neural networks (NNs), as an important part of deep learning architectures, can approximate complex nonlinear functions. It turns out that NNs have great potential for solving some intricate problems that cannot be described by analytical methods easily [2]. Research on NNs for optical communication systems has been increasingly popular and has started to be successfully used in many optical communication systems. In coherent optical orthogonal frequency division multiplexing systems, NNs have been tried to mitigate the nonlinear propagation effects [3,4]. Also, NNs have been implemented in intensity modulation and direct detection (IM/DD) systems to overcome both the linear and nonlinear distortions. To reach 50 Gb/s four level pulse amplitude modulations (PAM4) systems over 10 GHz devices, NN was used to optimize the equalization in the receiver side [5]. 32 GBaud and 40 Gbaud PAM8 IM/DD systems with NN equalizers were demonstrated [6,7]. NN has been considered to be a good solution for eliminating the channel distortion in other communication systems. NN-based equalizers for indoor optical wireless communications [8], 170 Mb/s visible light communications system [9] and modulation format identification in heterogeneous fiber-optic networks were proposed [10]. As a promising type of recurrent neural network, reservoir computing in all-optical implementation enables high-speed signal processing and can set the framework for a new generation of hardware for computing and future optical networks [11].

Although the use of NNs can often bring good results, the high computational complexity of NNs is a problem that cannot be ignored. Generally speaking, the computational complexity of simple multilayer perceptron NN-based equalizers is higher than that of the traditional linear feed-forward equalizer (FFE) and even the Volterra nonlinear equalizer [7,12], not to mention the NNs with more complex structures, such as long short-term memory NN or convolutional NN. The cost of training a NN is very high in terms of computational complexity and size of the training set, which might be not well-suited for some communication systems to perform real-time digital signal processing. Furthermore, there are dangers of overestimating the performance gain when applying NN in systems with pseudo-random bit sequences (PRBS) or with limited memory depths [13]. The use of PRBS may lead to overestimation of the NN performance. However, this issue is beyond the scope of this paper, so it is not discussed here. There is a trade-off between the accuracy and the number of training samples used in the training process for NNs [14]. Since it is believed that more data beats better algorithm [15], scientists in the field of artificial intelligence always use large-scale training data sets to train NNs. As a result, many efficient optimization algorithms have been proposed to ensure fast and stable convergence of minimizing the error function of the NNs models [16–18].

Many problems in many fields of science and engineering can be converted into the optimization problems of maximizing or minimizing objective functions by adjusting the parameters. Gradient-based optimization algorithms are the most commonly used optimization methods in these fields but not the exclusive tools of NN research. The optimization algorithms mainly include the first- and second-order optimization algorithms [19]. However, there are two main limitations in the second-order optimization algorithm, such as Newton's method and its variants. One limitation of the second-order optimization algorithms is that the cost function must be smooth, and the second derivatives are available or numerical approximation is achievable. Another limitation is that the Hessian matrix must be positive definite, and its dimension had better not be too large, taking the computational load into consideration [20]. Thus, the first-order gradient-based optimization algorithms, i.e., gradient descent and its variants are widely used in deep learning problems [21].

If the first partial derivatives of the cost function are available with respect to parameters, the gradient descent method is a very effective optimization method. To properly adjust the parameters, in every training iteration, the optimization algorithm calculates a gradient vector. Then the parameters is changed in the opposite direction of the gradient vector. It is worth noting that the commonly used tap estimation algorithms in time-domain equalizers (TDE) with the traditional structures, i.e., least mean square (LMS) and recursive least square algorithms, are based on first-order gradient of the cost function. Therefore, it is very possible to optimize the TDE with traditional structure by using gradient descent and its variants. In optical communications, especially short-reach optical communications, the distortion models are almost certain. The traditional TDE have been widely applied in optical communication systems and achieved good performance. Therefore, NNs are not always required in some communication systems with deterministic distortion models, taking the cost of training into consideration. In this paper, we propose and evaluate the tap estimation schemes for the equalizer with traditional structures in optical fiber communications using the optimization algorithms commonly used in the NNs. The experimental results show that adaptive moment estimation algorithm and batch gradient descent method perform well in the tap estimation of equalizer. In conclusion, the optimization algorithms of NNs are useful in the tap estimation of equalizer with traditional structures in optical communications.

The rest of this paper is organized as follows. In Section 2, we first review the mathematical model of FFE. Section 3 presents the principle of the proposed tap estimation schemes using optimization algorithms of NNs. The experimental setup is described in Section 4 and the detailed results and discussions are provided in Section 5. Finally, Section 6 concludes the paper.

2. Optimization Problems of Equalizers with Traditional Structures

Since linear transfer function can be generated by the FFE with the feature of easy implementation, it played a very significant role in compensating the channel impairments [22,23]. As depicted in Figure 1, the optimal tap coefficients are estimated by the optimization algorithms. As an application of stochastic gradient descent (SGD) method , LMS algorithm aims to minimize the current square error between the training sample y_t and output of the equalizer s_t, which can be expressed as [24]

$$\underset{\omega_i}{\text{minimize}}(\sum_{i=1}^{N} x_{j-i+1} \times \omega_i - y_t)^2 \tag{1}$$

where N is the number of taps and ω_i is the tap coefficient, for $i = 1, 2, ..., N$. In addition, x_j is the received signal.

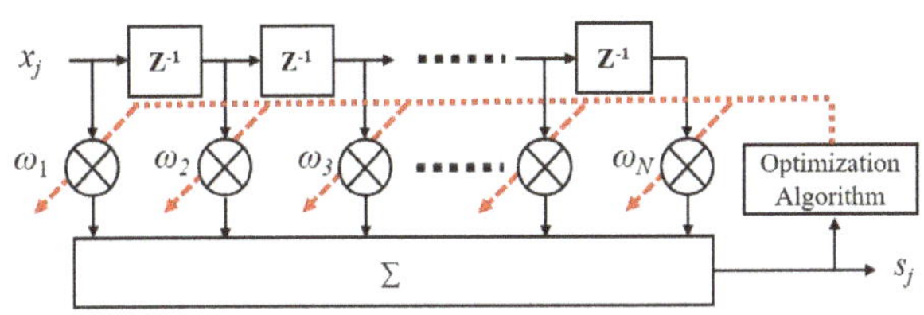

Figure 1. Schematic of an FFE with N taps.

3. Optimization Algorithms for Equalizers with Traditional Structures

3.1. BGD Method

There are three variants of gradient descent, including SGD, batch gradient descent (BGD) and mini-batch gradient descent (Mini-BGD) [21]. The main difference is that they use different number of training sample every training iteration. SGD updates the gradient of error function with respect to the only one training sample, while BGD uses all training samples to perform the parameters updating every iteration. To solve the problem of sharp increase in computation by using all samples every iteration in BGD, partial training samples are used in Mini-BGD every iteration. However, different from deep learning training, which requires large-scale training set, in practical communication systems, it is necessary to use as few training samples as possible. As a result, there is no need to adopt Mini-BGD but using BGD directly with a relatively small batch size. BGD method aims to minimize the mean square error (MSE) of all training samples, which can be expressed as

$$\underset{\omega_i}{\text{minimize}} \frac{1}{M-N+1} \sum_{j=N}^{M} (\sum_{i=1}^{N} x_{j-i+1} \times \omega_i - y_t)^2 \tag{2}$$

where M is the total number of training samples. BGD method using gradients for all training samples to perform just one update of tap coefficients and thus it is a global optimization algorithm. For better convergence, the first $N-1$ training samples are discarded [24]. Matrix form of Equation (2) can be express as

$$\underset{\omega}{\text{minimize}} \frac{1}{M-N+1} (\boldsymbol{R}\omega - \boldsymbol{Y})^T (\boldsymbol{R}\omega - \boldsymbol{Y}) \tag{3}$$

where $\omega = [\ \omega_1\ \ \omega_2\ \ \ldots\ \ \omega_N\ ]^T$ is the tap coefficient vector of the FFE and $Y = [\ y_N\ \ y_{N+1}\ \ \cdots\ \ y_M\ ]^T$ is the desired training vector. The $(M-N+1)$-by-N training matrix $\boldsymbol{R}$ can be expressed as

$$\boldsymbol{R} = \begin{bmatrix} x_N & x_{N-1} & \cdots & x_1 \\ x_{N+1} & x_N & \cdots & x_2 \\ \vdots & \vdots & \ddots & \vdots \\ x_M & x_{M-1} & \cdots & x_{M-N+1} \end{bmatrix} \tag{4}$$

where x_j are the received training samples, for $j = 1, 2, ..., M$. Gradient of MSE with respect with tap coefficient can be calculated as

$$\mathbf{g}(\omega) = \frac{2}{M-N+1}\boldsymbol{R}^T(\boldsymbol{R}\omega - \boldsymbol{Y}). \tag{5}$$

BGD method updates tap coefficient vector ω in the opposite direction of the gradient, which can be expressed as

$$\omega_t = \omega_{t-1} - \theta \times \mathbf{g}(\omega_{t-1}) \tag{6}$$

where θ is a positive step size and subscript t denotes t-th iteration.

NN researchers have long realized the fact that step size is one of the most difficult hyper-parameters to determine but it is critical to model performance and training costs. Then the following three adaptive step size optimization algorithms are successively proposed for NNs. These algorithms can also be used in tap estimation of equalizers with traditional structures.

3.2. AdaGrad

AdaGrad optimization algorithm scales the step sizes of all tap coefficients inversely proportional to cumulative squared gradient [16]. The cumulative squared gradient can be expressed as

$$\boldsymbol{r}_t = \boldsymbol{r}_{t-1} + \mathbf{g}(\omega_{t-1}) \odot \mathbf{g}(\omega_{t-1}) \tag{7}$$

where the cumulative squared gradient is initialized as N-by-1 zeros vector and $\odot$ denotes the Hadamard product. Gradient can be calculated from Equation (5) and tap coefficients are updated by

$$\omega_t = \omega_{t-1} - \frac{\theta}{\delta + \sqrt{\boldsymbol{r}_{t-1}}} \odot \mathbf{g}(\omega_{t-1}) \tag{8}$$

where a small value δ is used for numerical stability.

3.3. RMSProp

RMSprop is an unpublished, adaptive learning rate optimization algorithm [17]. By changing the cumulative gradient into an exponentially weighted moving average, RMSProp algorithm is derived from AdaGrad algorithm. Compared to AdaGrad, a new hyper-parameter ρ is added to control the scale of the moving average, which can be expressed as

$$\boldsymbol{r}_t = \rho \times \boldsymbol{r}_{t-1} + (1-\rho) \times \mathbf{g}(\omega_{t-1}) \odot \mathbf{g}(\omega_{t-1}). \tag{9}$$

Tap coefficients are also updated as by Equation (8).

3.4. Adam

Adaptive moment estimation (Adam), obtains individual adaptive step sizes for each tap coefficients from estimates of first and second moments of the gradients [18]. The biased first and second moment estimates $\mathbf{m}_t$ and $\mathbf{v}_t$ of $\mathbf{g}(\omega_t)$ are initialized as zeros vector, which are updated as

$$\mathbf{m}_t = \beta_1 \times \mathbf{m}_{t-1} + (1-\beta_1) \times \mathbf{g}(\omega_t), \tag{10}$$

$$\mathbf{v}_t = \beta_2 \times \mathbf{v}_{t-1} + (1-\beta_2) \times \mathbf{g}(\omega_t) \odot \mathbf{g}(\omega_t) \tag{11}$$

where β_1 and β_2 are set to 0.9 and 0.999, respectively. However, they are biased towards zero, especially during the first few steps. Bias-corrected first and second moment estimates counteract biases,

$$\hat{\mathbf{m}}_t = \mathbf{m}_t / \left(1-\beta_1^t\right), \tag{12}$$

$$\hat{\mathbf{v}}_t = \mathbf{v}_t / \left(1-\beta_2^t\right). \tag{13}$$

Tap coefficients are also updated by

$$\omega_t = \omega_{t-1} - \frac{\theta}{\delta + \sqrt{\hat{\mathbf{v}}_t}} \odot \hat{\mathbf{m}}_t. \tag{14}$$

4. Experimental Setups

The performance of the optimization algorithms of NNs that introduced to FFE are investigated by a 129-Gbit/s optical PAM8 system. The Figure 2 shows the experimental setups. At the transmitter, the input bits are first modulated to PAM8 symbols. After added 2000 training samples and 120 synchronized tokens, the digital PAM8 frames are uploaded into a digital-to-analog converter (DAC) with 86-GSa/s sampling rate and 16-GHz 3-dB bandwidth to generate electrical PAM8 frames. There are 82,360 PAM8 symbols per frames and the symbol rate of electrical PAM8 frames is 43 GBaud. A 40-Gbit/s electro-absorption integrated laser modulator (EML) modulates the electrical PAM8 frames to generate the optical PAM8 frames. Next, the generated optical PAM8 signals are launched into 2-km standard single-mode fiber (SSMF). At the receiver, received optical power (ROP) of the signals is adjusted by a variable optical attenuator (VOA). Then the received optical signals are converted into electrical signals by a photodiode (PD). The electrical signals are converted into digital signals by a real-time oscilloscope (RTO) with sampling rate of 80 GSa/s and 3-dB bandwidth of 36 GHz. Finally, off-line processing is implemented to deal with the digital signals, including re-sampling, synchronization, equalization using the FFE with optimization algorithms of NNs, post filter, maximum likelihood sequence detection (MLSD), PAM8 demodulation. The tap number of FFE is set to 101. After equalization, the high-frequency noise is amplified, which greatly degrades the performance of the system. So, a two-tap post filter is adopted to suppress the amplified high-frequency noise [25,26]. Furthermore, a known ISI is introduced by the post filter unavoidably, but it can be eliminated by MLSD based on Viterbi algorithm [27,28].

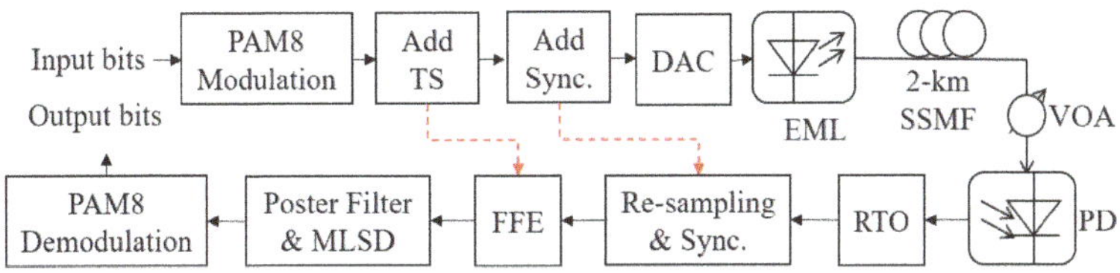

Figure 2. Block diagram of 129-Gbit/s optical PAM8 system. TS, training samples; DAC, digital-to-analog converter; EML, electro-absorption modulator integrated laser; SSMF, standard single-mode fiber; VOA, variable optical attenuator; PD, photodiode; RTO, real-time oscilloscope; Sync, Synchronization.

5. Results and Discussion

In this section, experiment results based on the setup described above are presented and the discussions are also provided. Figure 3 depicts the MSE curves of optimization algorithms versus iteration at ROP of −1 dBm after 2-km SSMF transmission. The iteration numbers of all these optimization algorithms are 120 times. It is clear that after 120 iteration, the MSE curves of using BGD method and Adam algorithm is lower than those of AdaGrad and RMSProp algorithms, which need more iterations to minimize the MSE. The MSE curve of BGD method drops rapidly and steadily because BGD method updates the tap coefficients at the gradient direction. Although the MSE curve of Adam algorithm fluctuates, it drops more quickly. The reason is that it does not update the tap coefficients at the gradient direction but it computes adaptive step size for different tap coefficients from estimates of first and second moments of gradients. It is obvious that tap estimation of FFE using Adam algorithm converges to a lower value of MSE than the other three algorithms. As can be seen from the insets, the diagrams of using BGD method and Adam algorithm are slightly clearer than those of using AdaGrad and RMSProp algorithms, which indicates that BGD method and Adam algorithm may be more effective in tap estimation of FFE.

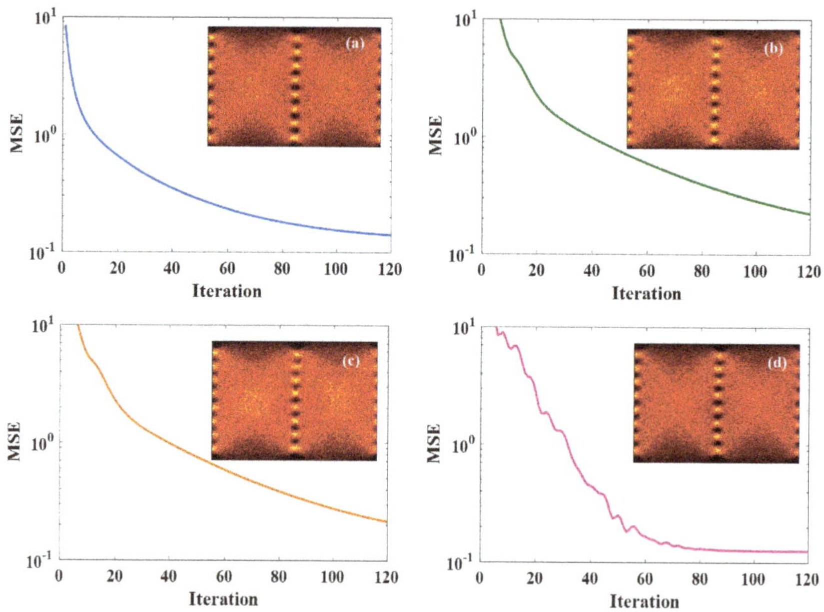

Figure 3. MSE curves of different optimization algorithms applied to 129-Gbit/s optical PAM8 system at ROP of −1 dBm after 2-km SSMF transmission. (**a**) BGD method; (**b**) AdaGrad algorithm; (**c**) RMSProp algorithm; and (**d**) Adam algorithm. Insets are eye diagrams of the equalized signals using the corresponding optimization algorithm.

The bit error rate (BER) performance of 129-Gbit/s PAM8 system versus ROPs at back-to-back (BTB) and 2-km transmission are shown in Figure 4, which can indicate the effectiveness of applying the optimization algorithms of NNs to tap estimation of traditional TDE. As shown in Figure 4a, after BTB transmission, 129-Gbit/s PAM8 system using the above four optimization algorithms have almost the same BER performance. At ROP of −5 dBm, the BER performance is below 7% forward error correction (FEC) limit. Moreover, after 2-km transmission, the BER performance of the system using AdaGrad and RMSProp algorithms are also almost same and below 7% FEC limit when the ROP is greater than or equal to −2 dBm. However, the BER performance of using BGD method is better than those of using AdaGrad and RMSProp algorithms. Although it is believed that AdaGrad and RMSProp algorithms are effective and practical in deep learning [29], for tap estimation of equalizer in the communication system, it seems that BGD method is better than them in terms of BER performance and computational complexity. The reason may be that these two algorithms are proposed to solve the sparse gradients of the NN and are not particularly superior in the traditional equalizer with simple structure [18]. It is empirically shown that Adam algorithm is usually better than other optimization algorithms in deep learning [18]. In this experiment, Adam also performs better than the other three optimization algorithms. The BER performance of using Adam algorithm is slightly better than that of BGD method and the ROP of using Adam algorithm is more than 1 dB lower than those of using AdaGrad and RMSProp algorithms at the 7% FEC limit. Therefore, the net rate of the system is 117 Gbit/s ($3 \times 43 \times$ 80,240/82,360/$(1+7\%) \approx$ 117 Gbit/s).

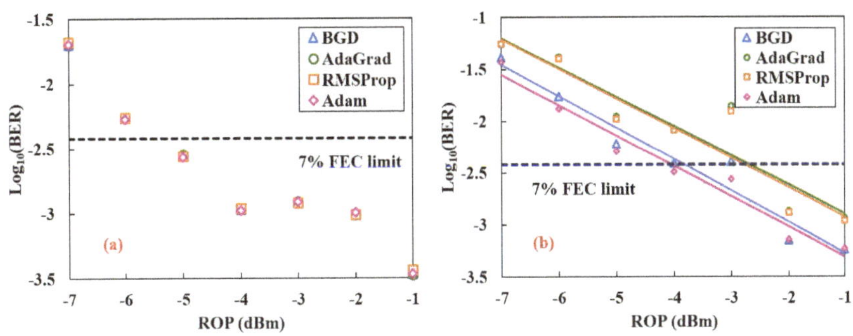

Figure 4. BER performance of 129-Gbit/s PAM8 system versus ROPs at BTB (**a**) and 2-km transmission (**b**) with FFE, post filter, and MLSD. FFE employs BGD (blue triangle), AdaGrad (green rhombus), RMSProp (orange square) and Adam (purple circle), respectively.

Next, the robustness of applying these algorithms to tap estimation of traditional TDE is going to be discussed. Just like LMS algorithm, trial and error is usually required to determine the effective step size of these optimization algorithms, to lead to satisfaction of fast and stable convergence of the tap estimation [30]. As a result, the range of effective step size plays an important role in the robustness of an optimization algorithm. In a general way, the wider the effective step size range of an optimization algorithm is, the better the robustness of the algorithm is. As shown in Figure 5, different optimization algorithms have different effective step sizes and ranges. In this experiment, for BGD method, AdaGrad, RMSProp, and Adam algorithms, the optimal step sizes are about 0.013, 0.3, 0.01 and 0.1, respectively. In general, when the step size is too large, they may not converge or even diverge; but when the step size is too small, they require a lot of iteration [29]. The effective ranges of step size of these four algorithms are respectively about 0.01, 0.87, 0.03 and 7.50. It is obvious that the effective step size range of BGD method is significantly narrower than that of the other three algorithms, because they are adaptive step size algorithms, which improve the robustness of the algorithms at the cost of increasing computational complexity. It is worth noting that Adam algorithm not only has the best BER performance, but also has the widest effective step size range, which means that it is more robust. Their effective ranges of step size in ascending order are as follows, BGD $<$ RMSProp $<$ AdaGrad $\ll$ Adam.

Finally, we analyze the computational complexity per iteration of the above optimization algorithms using in tap estimation. As shown in Table 1, BGD method has the lowest computational complexity, i.e., the minimum number of addition (Add.), multiplication (Mul.) operations and square root (Sqrt.) calculations. Although Adam algorithm has the highest computational complexity compared to the other three algorithms, the difference among them is not obvious. Their complexity in ascending order are as follows, BGD $<$ AdaGrad $<$ RMSProp $<$ Adam.

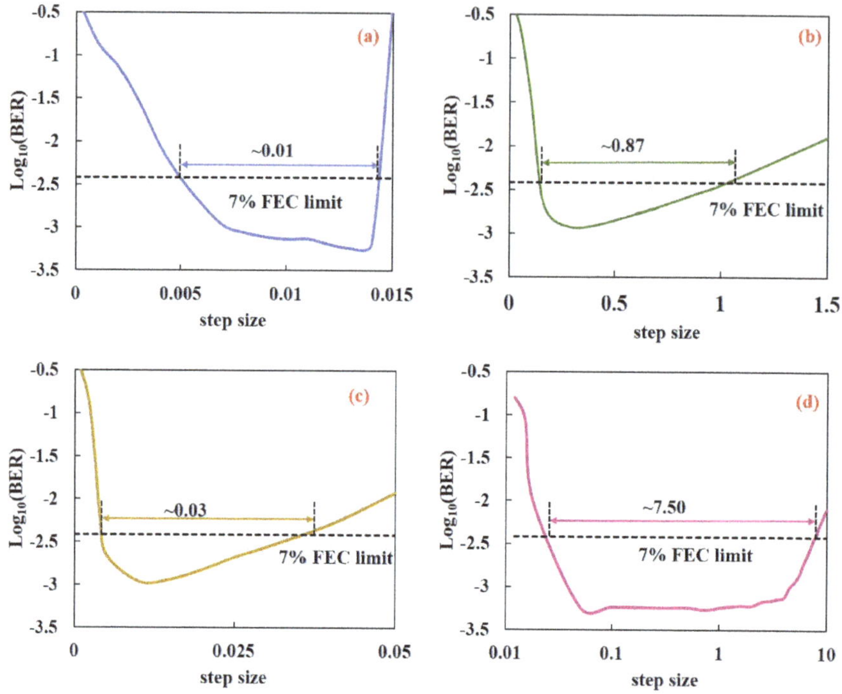

Figure 5. BER versus step size for 129-Gbit/s optical PAM8 system with different optimization algorithms at ROP of −1 dBm after 2-km SSMF transmission. (**a**) BGD method; (**b**) AdaGrad algorithm; (**c**) RMSprop algorithm; and (**d**) Adam algorithm.

Table 1. Computational complexity per iteration of optimization algorithms of NNs introduced to tap estimation of traditional equalizer.

	BGD	AdaGrad	RMSProp	Adam
Add.	$(2N-1)(M-N+1)$	$(2N-1)(M-N+1)+N$	$(2N-1)(M-N+1)+N$	$(2N-1)(M-N+1)+2N$
Mul.	$N[2(M-N+1)+1]$	$N[2(M-N+1)+3]$	$N[2(M-N+1)+5]$	$N[2(M-N+1)+9]$
Sqrt.	0	N	N	N

6. Conclusions

In this paper, we propose and evaluate the tap estimation schemes for the traditional TDE in optical fiber communications using the optimization algorithms commonly used in the NNs. The experimental results show that the optimization algorithms of NNs are also useful in tap estimation of optical communication system. BER performance of 129-Gbit/s PAM8 optical communication system adopting BGD method, AdaGrad, RMSProp, and Adam algorithms are all below 3.8×10^{-3}. It is also shown that Adam algorithm and BGD method perform better in the tap estimation of equalizer. Although Adam algorithm has the highest computational complexity compared to the other three algorithms, its performance is best and it is most robust with the effective step size range of ~7.50. BGD method performs better than AdaGrad and RMSProp algorithms and it is more straightforward to implement, but it is less robust with the effective step size range of ~0.01. In conclusion, the optimization algorithms of NNs are useful in the tap estimation of equalizer with traditional structures in optical communications.

Author Contributions: Conceptualization, H.W., J.Z., J.W., W.L., C.Y. and Z.L.; methodology, H.W., J.Z., J.W., W.L., C.Y. and Z.L.; software, H.W., J.Z. and Y.W.; formal analysis, H.W. and J.Z.; investigation, H.W., J.Z. and Y.W.; writing—original draft preparation, H.W., J.Z. and Y.W.; writing—review and editing, H.W., J.Z., W.L., C.Y. and Z.L.; funding acquisition, J.Z., W.L. and Z.L.

Funding: This research was funded by National Key R&D Program of China (2018YFB1801704); The Science and Technology Planning Project of Guangdong Province (2017B010123005, 2018B010114002); Local Innovation and Research Teams Project of Guangdong Pearl River Talents Program (2017BT01X121); National Science Foundation of China (NSFC) (61525502, 61975242); The Fundamental Research Funds for the Central Universities (21619309).

Conflicts of Interest: The authors declare no conflict of interest.

References

1. LeCun, Y.; Bengio, Y.; Hinton, G. Deep learning. *Nature* **2015**, *521*, 436–444. [CrossRef] [PubMed]
2. Schmidhuber, J. Deep learning in neural networks: An overview. *Neural Netw.* **2015**, *61*, 85–117. [CrossRef] [PubMed]
3. Jarajreh, M.A.; Giacoumidis, E.; Aldaya, I.; Le, S.T.; Tsokanos, A.; Ghassemlooy, Z.; Doran, N.J. Artificial neural network nonlinear equalizer for coherent optical OFDM. *IEEE Photonics Technol. Lett.* **2014**, *27*, 387–390. [CrossRef]
4. Ahmad, S.T.; Kumar, K.P. Radial basis function neural network nonlinear equalizer for 16-QAM coherent optical OFDM. *IEEE Photonics Technol. Lett.* **2016**, *28*, 2507–2510. [CrossRef]
5. Ye, C.; Zhang, D.; Huang, X.; Feng, H.; Zhang, K. Demonstration of 50Gbps IM/DD PAM4 PON over 10GHz class optics using neural network based nonlinear equalization. In Proceedings of the 2017 European Conference on Optical Communication (ECOC), Gothenburg, Sweden, 17–21 September 2017; pp. 1–3.
6. Gaiarin, S.; Pang, X.; Ozolins, O.; Jones, R.T.; Da Silva, E.P.; Schatz, R.; Westergren, U.; Popov, S.; Jacobsen, G.; Zibar, D. High speed PAM-8 optical interconnects with digital equalization based on neural network. In Proceedings of the 2016 Asia Communications and Photonics Conference (ACP), Wuhan, China, 2–5 November 2016; pp. 1–3.
7. Gou, P.; Yu, J. A nonlinear ANN equalizer with mini-batch gradient descent in 40Gbaud PAM-8 IM/DD system. *Opt. Fiber Technol.* **2018**, *46*, 113–117. [CrossRef]
8. Rajbhandari, S.; Ghassemlooy, Z.; Angelova, M. Effective denoising and adaptive equalization of indoor optical wireless channel with artificial light using the discrete wavelet transform and artificial neural network. *J. Light. Technol.* **2009**, *27*, 4493–4500. [CrossRef]
9. Haigh, P.A.; Ghassemlooy, Z.; Rajbhandari, S.; Papakonstantinou, I.; Popoola, W. Visible light communications: 170 Mb/s using an artificial neural network equalizer in a low bandwidth white light configuration. *J. Light. Technol.* **2014**, *32*, 1807–1813. [CrossRef]
10. Khan, F.N.; Zhou, Y.; Lau, A.P.T.; Lu, C. Modulation format identification in heterogeneous fiber-optic networks using artificial neural networks. *Opt. Express* **2012**, *20*, 12422–12431. [CrossRef]
11. Sorokina, M.; Sergeyev, S.; Turitsyn, S. Fiber echo state network analogue for high-bandwidth dual-quadrature signal processing. *Opt. Express* **2019**, *27*, 2387–2395. [CrossRef]
12. An, S.; Zhu, Q.; Li, J.; Ling, Y.; Su, Y. 112-Gb/s SSB 16-QAM signal transmission over 120-km SMF with direct detection using a MIMO-ANN nonlinear equalizer. *Opt. Express* **2019**, *27*, 12794–12805. [CrossRef]
13. Eriksson, T.A.; Bülow, H.; Leven, A. Applying neural networks in optical communication systems: Possible pitfalls. *IEEE Photonics Technol. Lett.* **2017**, *29*, 2091–2094. [CrossRef]
14. Bottou, L.; Bousquet, O. The tradeoffs of large scale learning. In Proceedings of the Neural Information Processing Systems Conference, Vancouver, BC, Canada, 3–8 December 2007; pp. 161–168.
15. Domingos, P.M. A few useful things to know about machine learning. *Commun. ACM* **2012**, *55*, 78–87. [CrossRef]
16. Duchi, J.; Hazan, E.; Singer, Y. Adaptive subgradient methods for online learning and stochastic optimization. *J. Mach. Learn. Res.* **2011**, *12*, 2121–2159.
17. Tieleman, T.; Hinton, G. Lecture 6.5-rmsprop: Divide the gradient by a running average of its recent magnitude. *COURSERA: Neural Networks Mach. Learn.* **2012**, *4*, 26–31.
18. Kingma, D.P.; Ba, J. Adam: A method for stochastic optimization. *arXiv* **2014**, arXiv:1412.6980
19. Battiti, R. First-and second-order methods for learning: between steepest descent and Newton's method. *Neural Comput.* **1992**, *4*, 141–166. [CrossRef]

20. Xiong, X.; De la Torre, F. Supervised descent method and its applications to face alignment. In Proceedings of the IEEE Conference on Computer Vision and Pattern Recognition, Portland, OR, USA, 23–28 June 2013; pp. 532–539.
21. Ruder, S. An overview of gradient descent optimization algorithms. *arXiv* **2016**, arXiv:1609.04747.
22. Nielsen, T.; Chandrasekhar, S. OFC 2004 workshop on optical and electronic mitigation of impairments. *J. Light. Technol.* **2005**, *23*, 131–142.
23. Watts, P.M.; Mikhailov, V.; Savory, S.; Bayvel, P.; Glick, M.; Lobel, M.; Christensen, B.; Kirkpatrick, P.; Shang, S.; Killey, R.I. Performance of single-mode fiber links using electronic feed-forward and decision feedback equalizers. *IEEE Photonics Technol. Lett.* **2005**, *17*, 2206–2208. [CrossRef]
24. Haykin, S.S. *Adaptive Filter Theory*; Pearson Education India: Noida, India, 2005.
25. Wang, H.; Zhou, J.; Li, F.; Liu, L.; Yu, C.; Yi, X.; Huang, X.; Liu, W.; Li, Z. Variable-step DD-FTN algorithm for PAM8-based short-reach optical interconnects. In Proceedings of the CLEO: Science and Innovations, San Jose, CA, USA, 5–10 May 2019; paper SW4O.3.
26. Zhou, J.; Qiao, Y.; Huang, X.; Yu, C.; Cheng, Q.; Tang, X.; Guo, M.; Liu, W.; Li, Z. Joint FDE and MLSD Algorithm for 56-Gbit/s Optical FTN-PAM4 System Using 10G-Class Optics. *J. Light. Technol.* **2019**, *37*, 3343–3350. [CrossRef]
27. Li, J.; Tipsuwannakul, E.; Eriksson, T.; Karlsson, M.; Andrekson, P.A. Approaching Nyquist limit in WDM systems by low-complexity receiver-side duobinary shaping. *J. Light. Technol.* **2012**, *30*, 1664–1676. [CrossRef]
28. Zhong, K.; Zhou, X.; Huo, J.; Yu, C.; Lu, C.; Lau, A.P.T. Digital signal processing for short-reach optical communications: A review of current technologies and future trends. *J. Light. Technol.* **2018**, *36*, 377–400. [CrossRef]
29. Goodfellow, I.; Bengio, Y.; Courville, A. *Deep Learning*; MIT Press: London, UK, 2016.
30. Marquardt, D.W. An algorithm for least-squares estimation of nonlinear parameters. *J. Soc. Ind. Appl. Math.* **1963**, *11*, 431–441. [CrossRef]

applied sciences

MDPI

Article

Mitigation of Nonlinear Impairments by Using Support Vector Machine and Nonlinear Volterra Equalizer

Rebekka Weixer *, Jonas Koch, Patrick Plany, Simon Ohlendorf and Stephan Pachnicke

Chair of Communications, Institute of Electrical Engineering and Information Technology, Kiel University, Kaiserstr. 2, 24143 Kiel, Germany

* Correspondence: rebekka.weixer@tf.uni-kiel.de

Received: 13 August 2019; Accepted: 4 September 2019; Published: 11 September 2019

Abstract: A support vector machine (SVM) based detection is applied to different equalization schemes for a data center interconnect link using coherent 64 GBd 64-QAM over 100 km standard single mode fiber (SSMF). Without any prior knowledge or heuristic assumptions, the SVM is able to learn and capture the transmission characteristics from only a short training data set. We show that, with the use of suitable kernel functions, the SVM can create nonlinear decision thresholds and reduce the errors caused by nonlinear phase noise (NLPN), laser phase noise, I/Q imbalances and so forth. In order to apply the SVM to 64-QAM we introduce a binary coding SVM, which provides a binary multiclass classification with reduced complexity. We investigate the performance of this SVM and show how it can improve the bit-error rate (BER) of the entire system. After 100 km the fiber-induced nonlinear penalty is reduced by 2 dB at a BER of 3.7×10^{-3}. Furthermore, we apply a nonlinear Volterra equalizer (NLVE), which is based on the nonlinear Volterra theory, as another method for mitigating nonlinear effects. The combination of SVM and NLVE reduces the large computational complexity of the NLVE and allows more accurate compensation of nonlinear transmission impairments.

Keywords: digital signal processing; support vector machines; BCSVM; nonlinear equalization; coherent detection

1. Introduction

The use of machine learning techniques in optical communication networks is currently a popular research topic [1]. Among the various algorithms for machine learning the support vector machine (SVM) can provide a powerful way of learning nonlinear functions. Besides noise, optical data transmission is also affected by linear and nonlinear impairments. Using coherent detection at the receiver, linear effects like chromatic dispersion can be successfully post-compensated by digital signal processing (DSP). Compensation can be done through a finite impulse response filter, also known as feed-forward equalizer (FFE). In case of long transmission distances, a separate electronic dispersion compensation (EDC) [2] is usually implemented, since otherwise too many coefficients for the adaptive FFE structure are required. With an increasing launch power, nonlinear effects additionally occur. For single-carrier transmission self-phase modulation (SPM), caused by the Kerr effect and nonlinear phase noise (NLPN), which results from the interaction between the amplified spontaneous emission (ASE) noise of inline optical amplifiers and SPM can be regarded as the most limiting nonlinear distortions [3]. These impairments cannot be compensated with conventional FFE structures. Previous approaches for the compensation of these nonlinear impairments focused on replacing the FFE by a nonlinear Volterra equalizer (NLVE) [4] or, if the fiber parameters are known, to replace the EDC with a digital backpropagation algorithm to compensate for linear and nonlinear effects simultaneously [5]. After using these methods, a signal detection with conventional linear decision thresholds takes place. Another approach for the compensation of nonlinear effects is an extended signal detection where the decision thresholds are adjusted to the disturbed constellations. In other words, the equalization problem is defined as a classification task. To solve this problem suitable algorithms such as expectation maximization (EM) [6,7], k-means algorithm (KMA) [8,9], neural network [10] or SVM [11] can be found in the large field of machine learning algorithms.

The advantage of extended signal detection by SVM is already emphasized in references [3,12–14]. In order to investigate exclusively the influence of nonlinearities such as NLPN or SPM, the influence of dispersion has been neglected deliberately in the past [3,12]. The absence of dispersion means that the interaction between dispersion and nonlinearities is not investigated. Thus, it should be examined whether these equalization techniques work equally well in dispersion influenced transmission.

In this paper we apply the SVM algorithm to a 64-QAM based coherent optical data center interconnect transmission system to mitigate nonlinear impairments after 100 km transmission distance, including the influence of dispersion. We numerically investigate, for the first time of our knowledge, the impact of different combinations of equalizer (FFE, NLVE) with various detection structures (SVM, KMA). Additionally, we show that the combination of SVM and NLVE can reduce the computational complexity of the NLVE and that this combination allows a more accurate compensation of the impairments that arise in an optical transmission system that is operated in the nonlinear regime.

2. Theoretical Analysis

2.1. Support Vector Machine

The support vector machine is a commonly used algorithm to classify data sets with binary output values. The method derived by Vladimir Vapnik is mainly based on the basics of statistical learning theory and applies the quadratic optimization problem to distinguish two classes in a feature space [15]. The training is done with a training set S of length N and consists of the input data x_i and binary classified data y_i:

$$S = (x_1, y_1), ..., (x_n, y_n), \qquad x_i \in \mathbb{R}^n, y_i \in \pm 1 \tag{1}$$

In order to separate the input features, a hyperplane $h(\mathbf{x})$ is calculated by a quadratic optimization problem. To allow more general decision surfaces, the input data is mapped to a higher dimensional feature space, that is, $\phi(\mathbf{x}) \in \mathbb{R}^m$, where the data is linearly separable. The SVM classifies an estimation according to

$$\hat{\mathbf{y}} = \text{sign}\{\mathbf{w}^T \phi(\mathbf{x}) + b\}, \tag{2}$$

where $\mathbf{w}$ and b are an orthogonal vector to $h(\mathbf{x})$ and a bias term, respectively, which are determined in a training process. The optimal hyperplane is found, if the margin—smallest distance between the hyperplane and any of the samples—is maximized [3,16].

The mentioned optimization problem distinguishes two classes whose feature vectors are located in an ideally delimitable area. If there are strong deviations of individual data points in the training set, for example, if a data point is located in the area of the contrary class, it is not possible to separate the data successfully. The problem can be solved by using a soft margin classifier. This enables a tolerance against data anomalies. For this purpose the optimization problem is extended by an error term including a weighting coefficient C and slack variables ξ [3]

$$\min_{\mathbf{w},b,\xi} \frac{1}{2}||w||^2 + C\sum_{i=0}^{N} \xi_i \tag{3}$$

under constraints

$$y_i(\mathbf{w}^T \phi(\mathbf{x}) + b \geq 1 - \xi_i) \tag{4}$$

$$\xi_i \geq 0, \quad i = 1, ..., N. \tag{5}$$

The basic algorithm of the SVM has been formulated in terms of scalar products in the feature space F. According to Mercer's theorem the intensive calculation in the higher dimensional space can be significantly reduced by a suitable kernel function $K(\mathbf{x}, \mathbf{x}_i) = \phi(\mathbf{x}) \cdot \phi(\mathbf{x}_i)$ [16]. With the use of different kernels, diverse problems can be solved, which opens up a wide variety of SVM learning machines. In case of M-QAM transmission, the radial base function (RBF) is the most suitable kernel and is defined by

$$K(\mathbf{x}, \mathbf{x}_i) = \exp(-\gamma||\mathbf{x} - \mathbf{x}_i||^2) \tag{6}$$

with the kernel parameter $\gamma > 0$ [16].

SVM-based Detection

For coherent optical communications systems, the modulation format M-QAM is usually selected. To be able to process the signals at the receiver by using an SVM, each cluster of the signal constellation represents one class, for example, 16-QAM consists of 16 classes. Since the SVM is fundamentally a binary classifier, an extension of the SVM structure is required. Various methods have been proposed for combining multiple binary SVMs in order to build a multi-class SVM. Common methods to extend the SVM are the one-vs-one (OVO), one-vs-all (OVA) and binary-coding SVM (BCSVM) or also called M-ary SVM [15]. The OVO principle is based on the comparison of two different classes from the entire set of all classes. If a data set contains N different classes, then $N(N-1)/2$ different binary SVMs are trained. For the concluding decision a voting procedure is used and the class with the highest vote is detected. In the OVA scheme the data of one class is separated from the complete data of all remaining $N-1$ classes. Thus, N SVMs are trained. At the end, the class with the highest vote is detected. For communication systems, the BCSVM is the most appropriate choice. The symbols or classes are already labelled in binary format, enabling each individual bit to be modelled with one conventional SVM. For M-QAM $\log_2(M)$ SVMs are required.

Figure 1a shows the principle of the BCSVM for 16-QAM. The respective SVMs are color-coded with the corresponding bits. Figure 1b illustrates the processing structure. The complex input vector $\mathbf{x}_{rx}$ contains the received symbols divided into real and imaginary parts. Training is done according to the transmitted bit sequence $\mathbf{y}$. The separating hyperplane of each SVM is determined by quadratic programming during training [3].

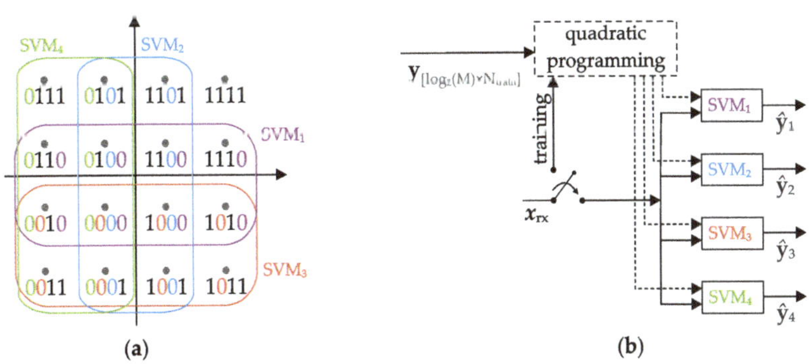

Figure 1. Binary Coding Support Vector Machine (BCSVM) nonlinear classification using four support vector machines (SVMs) [3]: (**a**) Coding and classification scheme for BCSVM based detection and (**b**) the processing structure for BCSVM used for 16-QAM signal detection.

The output of the binary SVM array is an estimation of the transmitted bit sequence $\hat{\mathbf{y}} = [\hat{\mathbf{y}}_1, \hat{\mathbf{y}}_2, \hat{\mathbf{y}}_3, \hat{\mathbf{y}}_4]$. Consequently, the received signal is classified and demodulated at the same time.

Figure 2 exemplarily shows an iteration of the training process of the presented methods. The data points of the two opposite classes are colored red or blue. It can be seen, that the OVA and BCSVM method, in contrast to the OVO, take into account all data points in each iteration step. This may result in a significant computational complexity, if too much training data is used. However, the OVO method requires significantly more iterations steps than the other methods [17].

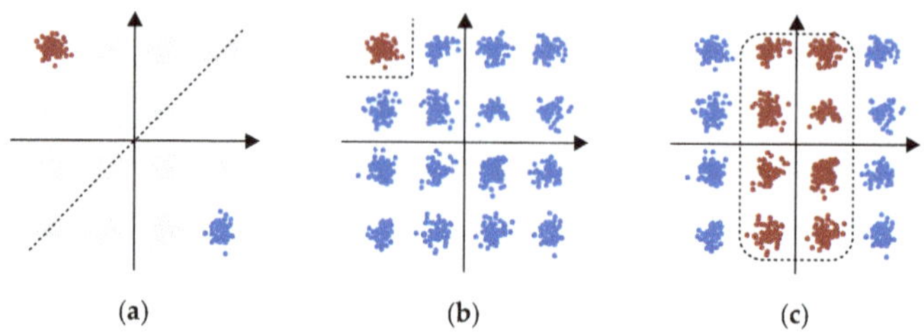

Figure 2. Illustration of one iteration during training for (**a**) OVO, (**b**) OVA and (**c**) BCSVM methods in case of 16-QAM transmission. The opposite classes are marked in red and blue and the corresponding hyperplane is indicated by the dashed line.

During the training process it is necessary to adapt certain parameters of the optimization problem to the characteristics of the input data. The aim is to avoid over- or underfitted systems, which leads to a significantly reduced classification accuracy [12]. The adaptation and verification is implemented using the two optimization algorithms Grid-Search and Cross Validation [12,18]. In the optimization process, 70% of the training data is used for training and 30% for validation.

2.2. Nonlinear Volterra Equalizer

The principle of the NLVE is based on the theory of the Volterra series, which is an important tool for the analysis of nonlinear systems and provides a complete description of the channel nonlinearity [19,20]. The realization can be done either in the frequency domain or entirely in the time domain, as will be shown here. A general discrete Volterra filter input-output relation is given by [4]

$$y_n = \sum_{v=0}^{N_1-1} e_v x_{n-v} + \sum_{v=0}^{N_2-1}\sum_{l=v}^{N_2-1} e_{v,l} x_{n-v} x_{n-l} + \sum_{v=0}^{N_3-1}\sum_{l=v}^{N_3-1}\sum_{m=0}^{N_3-1} e_{v,l,m} x_{n-v} x_{n-l} x^*_{n-m}, \tag{7}$$

where x_n and y_n are the complex-valued filter input and output of the equalizer at the time index n, N_i is the memory length of the i-th order and e_v, $e_{v,l}$, $e_{v,l,m}$ are the equalizer coefficients. The first term of Equation (7) represents a linear filter, whereas the others are nonlinear.

The coefficients can be estimated using the minimum mean square error (MMSE) criterion. The dimension of the model grows rapidly, as can be seen from the total number of coefficients given by

$$N_t = N_1 + N_2(N_2+1)/2 + N_3^2(N_3+1)/2. \tag{8}$$

According to Equation (7) we choose the notation NLVE[N_1,N_2,N_3] as full description of the Volterra filter. In this case N_1-N_3 represents the memory length of the 1st–3rd order of the NLVE.

3. Simulation Setup

The proposed techniques are subsequently thoroughly evaluated in numerical simulations for a 64 GBd 64-QAM system. For simulation purposes we will initially restrict ourselves to a single-polarization system but it can be extended straight forward to a dual-polarization system. A schematic of the general setup is given in Figure 3. A 2^{16} randomly generated bit sequence, using the MATLAB R2018a (9.4.0.813654) rand function, is mapped to the 64-QAM symbols. The digital to analog conversion is modelled as a root-raised cosine pulse shaping filter with roll-off factor $\beta = 0.3$. The symbols are modulated on the carrier (wavelength $\lambda_c = 1550$ nm) via an I/Q MZ-modulator.

The linewidth of the laser is set to zero. The modulated optical signal is coupled into the fiber after it is amplified by the erbium doped fiber amplifier (EDFA) with a noise figure (NF) of 5 dB.

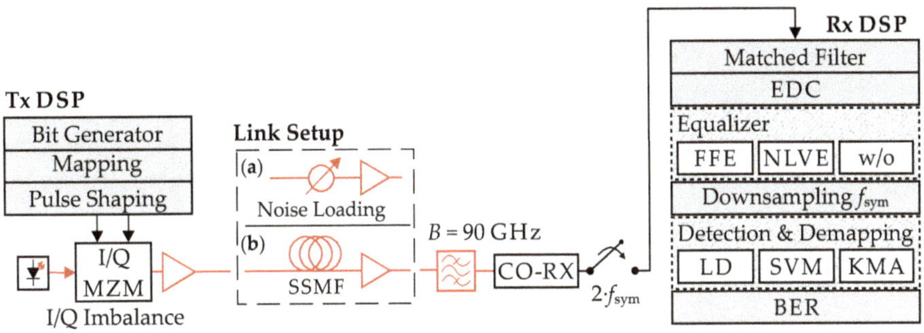

Figure 3. Simulation setup of the 64 Gbd 64-QAM single-polarization coherent optical simulation system including two different setups for the link. By using the setup (**a**) a B2B transmission with noise loading is examined. The setup (**b**) consists of a 100 km SSMF transmission with subsequent electronic dispersion compensation (EDC) to investigate a dispersion uncompensated link.

In order to investigate the performance of the enhanced detection algorithms, two types of communication systems are modeled. The link setup (**a**) is used to test a back-to-back (B2B) scenario, that is, no transmission link was simulated. The setup (**b**) is used to examine a dispersion uncompensated link, where the dispersion is compensated by DSP at the end of the transmission. The parameters for the SSMF are given by the attenuation coefficient $\alpha = 0.2$ dB/km, the dispersion coefficient $D = 17$ ps/(nm·km), dispersion slope $S = 0.06$ ps/(nm^2·km) and the nonlinear coefficient $\gamma = 1.3\,(\mathrm{W{\cdot}km})^{-1}$. For a complete compensation of span loss an EDFA (NF = 5 dB) is applied. After transmission a Gaussian optical filter with 90 GHz bandwidth is used to reduce ASE noise. The received signal is detected by a coherent receiver and downsampled to 128 GS/s. After matched filtering an ideal EDC is used to compensate for dispersion. After the equalization stage, which consists of either an FFE, an NLVE or no equalizer at all (w/o), the signal is downsampled to symbol frequency and detected. Detection and demodulation is performed either linear by using conventional linear decision thresholds and demapping, here called linear detection (LD), or by machine learning algorithms such as SVM or KMA [8,9]. System performance is evaluated by BER. The hard-decision forward error correction (HD-FEC) limit is assumed to be 3.7×10^{-3}. We examine the suitability of the SVM as a classifier and combine the mentioned equalizer schemes with the SVM to achieve the maximum gain of the machine learning algorithm.

Since more coefficients require more training symbols, increasing the number of coefficients without adjusting the number of training symbols might decrease the performance. Thus, for a correct adjustment of the Volterra equalizer it is necessary to determine the optimal number of coefficients and training symbols. For the further investigations the training length of 2048 symbols and memory lengths of NLVE[4,2,5] was determined after optimization.

4. Results and Discussion

Initially the behavior of SVM against I/Q imbalances was examined. In an I/Q modulator, the ideal phase shift between the I- and Q-branch is 90°. Due to physical imperfections of the system components and the non-perfect tuning of the $\pi/2$ phase shift, amplitudes and phase mismatches may occur. These I/Q imbalances may considerably disturb the signal constellation [21]. We investigate the I/Q imbalances in a B2B scenario according to Figure 3a at an optical signal-to-noise ratio (OSNR) of 28 dB, where the signal is disturbed at the transmitter side. The amplitude mismatch is set to 0.125 and the phase mismatch is varied between 0° and 30°. To cope with these imperfections, we examine the performance of the BCSVM and the OVA-SVM and compare them to LD. In order to compare the SVM with other enhanced detection techniques, the KMA is added to this comparison. The training length of the respective SVMs and KMA is set to 1024 symbols. Moreover, the number of iterations for KMA is set to 5.

Figure 4 shows the performance of the various detection methods depending on the transmitter I/Q imbalances. As expected, detection by machine learning algorithms is more robust against I/Q imbalances compared to LD. For low phase mismatch, the performance of the two enhanced detection techniques seems similar. However, above 12° phase mismatch the KMA's performance rapidly deteriorates. For SVM a decline in performance can be observed above 20 ° phase mismatch. Compared with OVA-SVM, the BCSM achieves slightly better performance, which is in the range of 1×10^{-4}.

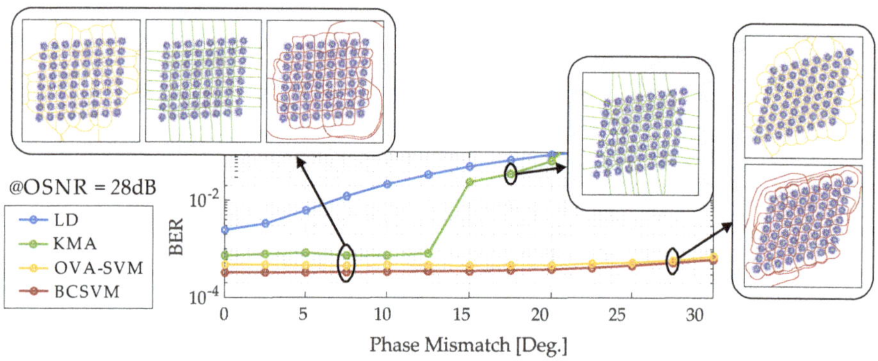

Figure 4. Simulations results in case of transmitter I/Q imbalances. BER vs. phase mismatch for an amplitude mismatch of 0.125 and 28 dB OSNR.

It should be mentioned that SVM and KMA are two completely different procedures. The SVM has already been introduced as classification algorithm in Section 2.1. The KMA, in contrast, belongs to a cluster-based detection. The training of KMA is iterative and unsupervised, while the training of the SVM is supervised. The KMA is initialized with the centers of the cluster. Therefore, it is necessary to know how many clusters are present and where the centers are approximately expected. If the actual cluster is too far away from the expected cluster, the KMA is no longer able to separate the clusters correctly. This can be seen for example in the constellation of the KMA at 17.5 ° phase mismatch, where the field at the top right has been assigned to about two full constellation points. Although the centers of the clusters are updated in each iteration. This effect may also occur in case of a phase rotation induced by SPM. Furthermore, the KMA is only a linear algorithm in essence, while the SVM is a nonlinear classifier due to the usage of kernels. Accordingly, the KMA is unsuitable for highly complex and nonlinear data distributions and is therefore no longer used as comparison in the following investigations.

Regarding the visualization of the decision thresholds, the different working principles of the algorithms can be observed. Based on the RBF kernel, the SVM calculates significantly rounder and softer decision thresholds than the KMA. Additionally, a difference between the multi-class methods of the SVM can be seen. Therefore, we would like to point out at this point that besides the selection of the kernel also the choice of the SVM multi-class method may have a more or less significant influence on the results.

Next, we include the fiber in our simulations. To evaluate the ability of the SVM to compensate nonlinear impairments in the 100 km setup for different launch powers, we compare the nonlinear detection by SVM with an FFE and an NLVE. To distort the 64-QAM constellation we set the modulation depth of the modulator to $m = V_{pp}/V_{\pi} = 2.2$ and generated an I/Q imbalance with 5% phase deviation from 90°. The number of training symbols for SVM is set to 1024.

The BER as a function of the launch power after 100 km dispersion uncompensated transmission is shown in Figure 5. The launch power of the 64-QAM signal ranges from −6 to +12 dBm. We investigate different combinations of equalizers and detection techniques. Figure 5a first presents the results for FFE[1] and NLVE[4,2,5] in conjunction with LD. In addition, the red curve shows a detection based on SVM only without any previously inserted equalizer. It can be seen that a nonlinear detection with SVM only is already quite powerful. Here, the lowest BER is achieved by SVM at 3 dBm launch power, which is about six times lower than the BER using the FFE. Up to 4 dBm the best results can be achieved with SVM detection. Above 4 dBm nonlinear effects dominate and the optimally configured NLVE shows the best performance while the SVM is not as good as the NLVE but still better than the FFE. The launch power to stay below HD-FEC can be increased by 2 dB, if NLVE[4,2,5] is used and by 1 dB if the SVM is used compared to FFE[1]. If an FFE[1] or NLVE[4,2,5] is now added before the SVM, the overall system performance can be improved significantly, as shown in Figure 5b,c. Especially the combination of NLVE[4,2,5] and SVM further minimizes the BER significantly as can be seen in Figure 5c at 3 dBm launch power, where the BER is reduced from 7.7×10^{-5} to 3.1×10^{-6} by SVM.

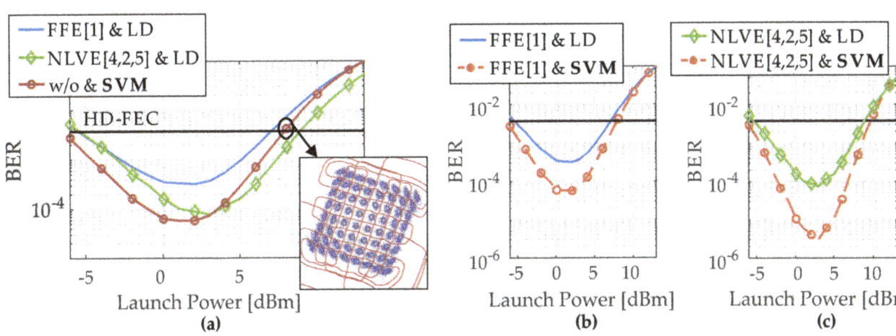

Figure 5. BER as a function of the launch power at 100 km dispersion unmanaged transmission: (**a**) shows equalization by FFE[1] and NLVE[4,2,5] combined with LD and only SVM detection. (**b**) shows the combined structure FFE[1] and SVM and (**c**) shows the combination NLVE[4,2,5] with SVM.

The optimum setting for the NLVE is given by NLVE[4,2,5]. So, the total number of NLVE coefficients sums up to $N_t = 82$, according to Equation (8). The majority of coefficients belongs to the third order of the NLVE. Therefore, in our further investigations we have reduced the number of delay elements in the third order to $N_3 = 3$. Consequently, the number of coefficients is decreased from 75 to 18 (74%). Figure 6 shows the obtained BER as a function of the launch power for the optimal and reduced NLVE. The SVM is trained with 1024 and the NLVE with 2048 symbols. It can be seen that further reducing the coefficients of the NLVE leads to a decline of the overall system performance. To stay below the HD-FEC, the launch power is reduced by 1 dB in case of NLVE[3,2,3] and LD compared to the NLVE[4,2,5] and LD. Furthermore, NLVE[4,2,3] is continuously worse than a detection by SVM only. By combining the reduced NLVE[4,2,3] and the SVM, it can be observed that better results are achieved compared to the optimally adjusted NLVE[4,2,5] and LD. At 3 dBm launch power, the BER of NLVE[4,2,3] and LD (green dashed line) is 1.2×10^{-4} and can be reduced to 1×10^{-5} by SVM (orange dashed line).

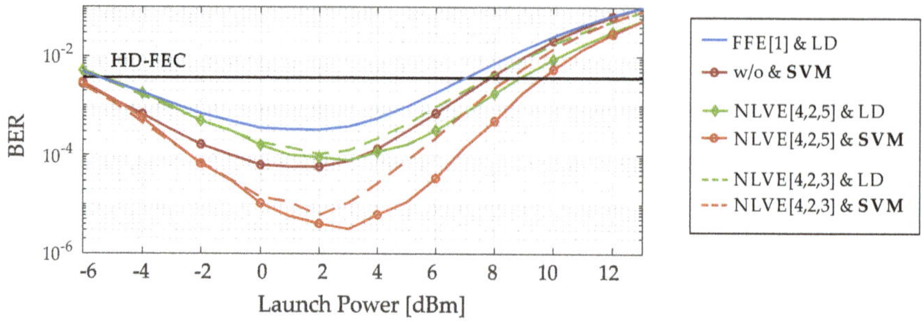

Figure 6. BER as a function of the launch power at 100 km for NLVE equalization with optimal and reduced number of coefficients in combination with SVM based detection.

Finally, we examined the impact of the number of training symbols on the performance of the NLVE and the SVM. The obtained results are presented in Figure 7 including the investigations for the NLVE with optimal and reduced coefficients in combination with LD and SVM based nonlinear detection plus a single detection only by SVM. We increased the number of training symbols from 512 to 3584 at 5 dBm launch power. The main improvement is observed after an increase from 512 to 1024 symbols for all investigated structures.

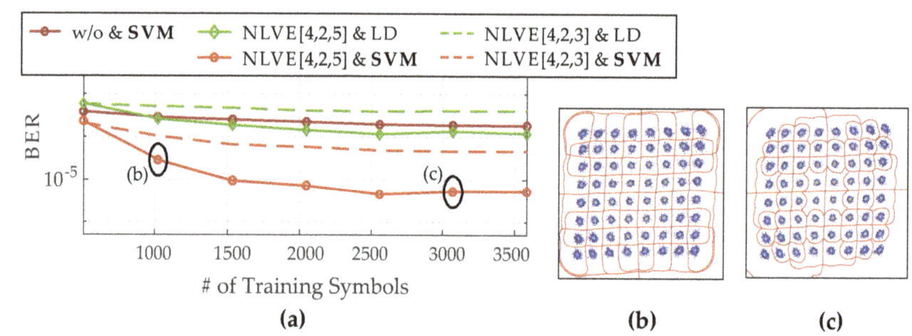

Figure 7. (**a**) Shows the BER as function of the number of training symbols for 100 km dispersion unmanaged transmission at 5 dBm launch power. (**b**) Shows the the corresponding constellation diagram for NLVE[4,2,5] & SVM trained with 1024 symbols and (**c**) Shows the constellation diagram for NLVE[4,2,5] & SVM trained with 3072 symbols.

The training of the SVM is based on the classes that are included in the classification task. To ensure that the SVM can learn and capture link properties from ony a small amount of training data [12], it is important, that besides a sufficient number of training symbols all classes are uniformly distributed in the training set. For example, with the amount of 512 training symbols and 64 different classes it is not guaranteed that each class is included in the training data, if a randomly generated training sequence is used. Concerning the NLVE, the training is based on the amount of inter symbol interference which is independent on the training data itself. Here, a certain number of symbols is necessary to estimate the coefficients correctly. In case of the NLVE[4,2,5], the training length of 512 is not sufficient to determine the coefficients correctly. However, if a certain number of training symbols is used, the channel estimation of the NLVE can improve its performance barely, even if more training data is used as it can be seen for the reduced NLVE[4,2,3]. While the plain NLVE and SVM structures saturated fast, the results with combined NLVE and SVM based detection are quite remarkable.

5. Conclusions

In this paper we compared and combined nonlinear detection by SVM with post-compensation techniques for the mitigation of nonlinearities regarding their performance and computational complexity. Unlike clustering and classification algorithms like EM or KMA, the SVM does not require any prior knowledge of the modulation format. We have shown that by combining NLVE and SVM based detection it is possible to improve the overall system performance for 64 GBd 64-QAM coherent transmission over 100 km. For example, at 3 dBm launch power, the BER is reduced from 7.7×10^{-5} to 3.1×10^{-6} by SVM. It is well known that nonlinear equalization using an NLVE is computationally quite complex, so a trade-off between complexity and performance is often required. Therefore, the performance of a reduced NLVE was evaluated and the obtained results have shown, that by adding an SVM it is possible to reduce the number of coefficients by 74% while maintaining or improving the overall system performance. The SVM classification approach provides a way to cluster datasets without prior knowledge of the channel characteristics or the modulation format. Based on the previous studies and discussions, we strongly believe that in context of coherent optical transmission systems, an enhanced detection by using SVM and its methods should be further investigated. So far, we mainly examined a single channel and single-polarization. Other effects among multiple channels and polarization effects will be taken into account in future studies.

Author Contributions: Conceptualization, R.W., J.K., P.P. and S.P. developed the concept; formal analysis, R.W. and J.K.; designed and performed the simulations and analyzed the data; writing—original draft preparation, R.W. and P.P.; writing—review and editing, J.K., S.O. and S.P.

Funding: This research received no external funding.

Acknowledgments: We gratefully acknowledge the support of NVIDIA Coorporation for the donation of the Titan Xp GPU used for this research.

Conflicts of Interest: The authors declare no conflict of interest.

Abbreviations

The following abbreviations are used in this manuscript:

ASE	Amplified Spontaneous Emission
BCSVM	Binary Coding Support Vector Machine
BER	Bit Error Ratio
B2B	Back-to-Back
DSP	Digital Signal Processing
EDC	Electronic Dispersion Compensation
EDFA	Erbium Doped Fiber Amplifier
HD-FEC	Hard-Decision Forward Error Correction
KMA	K-Means Algorithm
LD	Linear Detection
MMSE	Minimum Mean Square Error
NF	Noise Figure
NLPN	Nonlinear Phase Noise
OVA	One versus All
OVO	One versus One
OSNR	Optical Signal-to-Noise Ratio
QAM	Quadrature Amplitude Modulation
RBF	Radial Basis Function
SSMF	Standard Single Mode Fiber
SVM	Support Vector Machine

References

1. Musumeci, F. An overview on application of machine learning techniques in optical networks. *IEEE Commun. Surv. Tutor.* **2018**, *21*, 1383–1408. [CrossRef]
2. Bülow, H.; Buchali, F.; Klekamp, K. Electronic Dispersion Compensation. *J. Light. Technol.* **2008**, *26*, 158–167. [CrossRef]
3. Li, M.; Yu, S.; Yang, J.; Chen, Z.; Han, Y.; Gu, W. Nonparameter nonlinear phase noise mitigation by using M-ary support vector machine for coherent optical systems. *IEEE Photonics J.* **2013**, *5*, 7800312–7800312. [CrossRef]
4. Han, X.; Chen, C. Nonlinear filter based decision feedback equalizer for optical communication systems. *Opt. Express* **2014**, *22*, 8712–8719. [CrossRef] [PubMed]
5. Ip, E.; Kahn, J.M. Compensation of dispersion and nonlinear impairments using digital backpropagation. *J. Light. Technol.* **2008**, *262*, 3416–3425. [CrossRef]
6. Moon, T.K. The expectation-maximization algorithm. *IEEE Signal Process. Mag.* **1996**, *13*, 47–60. [CrossRef]
7. Zibar, D.; Winther, O.; Franceschi, N.; Borkowski, R.; Caballero, A.; Arlunno, V.; Schmidt, M.; Gonzales, N.; Mao, B.; Ye, Y.; et al. Nonlinear impairment compensation using expectation maximization for dispersion managed and unmanaged PDM 16-QAM transmission. *Opt. Express* **2012**, *20*, B181–B196. [CrossRef]
8. Aggarwal, C.C. A human-computer interactive method for projected clustering. *IEEE Trans. Knowl. Data Eng.* **2004**, *16*, 448–460. [CrossRef]
9. Dong, J.; Qi, M. K-means optimization algorithm for solving clustering problem. In Proceedings of the 2009 Second International Workshop on Knowledge Discovery and Data Mining, Moscow, Russia, 23–25 January 2009; pp. 52–55.
10. Chuang, C.-Y.; Liu, L.-C.; Wei, C.-C.; Liu, J.-J.; Henrickson, L.; Huang, W.-J.; Wang, C.-L.; Chen, Y.-K.; Chen, J. Convolutional neural network based nonlinear classifier for 112-Gbps high speed optical link. In Proceedings of the Optical Fiber Communications Conference (OFC), San Diego, CA, USA, 11–15 March 2018.
11. Bishop, C.M. Sparse kernel machines. In *Pattern Recognition and Machine Learning*; Springer: Berlin/Heidelberg, Germany, 2006; pp. 352–359.
12. Wang, D.; Zhang, M.; Cai, Z.; Cui, Y.; Li, Z.; Han, H.; Fu, M.; Luo, B. Combatting nonlinear phase noise in coherent optical systems with an optimized decision processor based on machine learning. *Opt. Commun.* **2016**, *369*, 199–208. [CrossRef]
13. Han, Y.; Yu, S.; Li, M.; Yan, J.; Gu, W. An SVM-based detection for coherent optical APSK systems with nonlinear phase noise. *IEEE Photonics J.* **2014**, *6*, 1–10. [CrossRef]
14. Lau, A.P.T.; Kahn, J.M. Signal design and detection in presence of nonlinear phase noise. *J. Light. Technol.* **2007**, *25*, 3008–3016. [CrossRef]
15. Sebald, D.J.; Bucklew, J.A. Support vector machines and the multiple hypothesis test problem. *IEEE Trans. Signal Process.* **2001**, *49*, 2865–2872. [CrossRef]
16. Schölkopf, B. *Support Vector Learning*; Technical University of Berlin: Berlin, Germany, 1997.
17. Zhibin, L.; Lianwen, J. LATTICESVM—A new method for multi-class Support Vector machines. In Proceedings of the 2008 IEEE International Joint Conference on Neural Networks (IEEE World Congress on Computational Intelligence), Hong Kong, China, 1–8 June 2008; pp. 727–733.
18. Hsu, C.-W.; Chang, C.-C.; Lin; C.-J. A practical guide to support vector classification. *BJU Int.* **2003**, *101*, 1396–1400.
19. Gao, Y.; Zhang, F.; Dou, L.; Chen, Z.; Xu, A. Intra-channel nonlinearities mitigation in pseudo-linear coherent QPSK transmission systems via nonlinear electrical equalizer. *Opt. Commun.* **2009**, *282*, 2421–2425. [CrossRef]
20. Peddanarappagari, K.V.; Brandt-Pearce, M. Volterra series transfer function of single-mode fibers. *J. Light. Technol.* **1997**, *15*, 2232–2241. [CrossRef]
21. Petrou, C.; Vgenis, A.; Roudas, I.; Raptis, L. Quadrature imbalance compensation algorithms for coherent PDM QPSK systems. *IEEE Photonics Technol. Lett.* **2009**, *21*, 1876–1878. [CrossRef]

applied sciences

MDPI

Article

Histogram Based Clustering for Nonlinear Compensation in Long Reach Coherent Passive Optical Networks

Ivan Aldaya [1,*], Elias Giacoumidis [2], Geraldo de Oliveira [1], Jinlong Wei [3], Julián Leonel Pita [4], Jorge Diego Marconi [5], Eric Alberto Mello Fagotto [6], Liam Barry [2] and Marcelo Luis Francisco Abbade [1]

1 Campus of São João da Boa Vista, State University of São Paulo, São João da Boa Vista 13876-750, Brazil
2 Radio and Optical Communications Lab, Dublin City University, Dublin, Ireland
3 European Research Center, Huawei Technologies Duesseldorf GmbH, 40549 Munich, Germany
4 School of Electrical Engineering, State university of Campinas, Campinas 13083-852, Brazil
5 Engineering, Modeling and Applied Sociology Center, Universidade Federal do ABC, São Paulo 09210-580, Brazil
6 School of Electrical Engineering, Pontifical Catholic University of Campinas, Campinas 13087-571, Brazil
* Correspondence: ivan.a.aldaya@ieee.org

Received: 6 September 2019; Accepted: 5 December 2019; Published: 23 December 2019

Abstract: In order to meet the increasing capacity requirements, network operators are extending their optical infrastructure closer to the end-user while making more efficient use of the resources. In this context, long reach passive optical networks (LR-PONs) are attracting increasing attention.Coherent LR-PONs based on high speed digital signal processors represent a high potential alternative because, alongside with the inherent mixing gain and the possibility of amplitude and phase diversity formats, they pave the way to compensate linear impairments in a more efficient way than in traditional direct detection systems. The performance of coherent LR-PONs is then limited by the combined effect of noise and nonlinear distortion. The noise is particularly critical in single channel systems where, in addition to the the elevated fibre loss, the splitting losses should be considered. In such systems, Kerr induced self-phase modulation emerges as the main limitation to the maximum capacity. In this work, we propose a novel clustering algorithm, denominated histogram based clustering (HBC), that employs the spatial density of the points of a 2D histogram to identify the borders of high density areas to classify nonlinearly distorted noisy constellations. Simulation results reveal that for a 100 km long LR-PON with a 1:64 splitting ratio, at optimum power levels, HBC presents a Q-factor 0.57 dB higher than maximum likelihood and 0.21 dB higher than k-means. In terms of nonlinear tolerance, at a BER of 2×10^{-3}, our method achieves a gain of $\sim$2.5 dB and $\sim$1.25 dB over maximum likelihood and k-means, respectively. Numerical results also show that the proposed method can operate over blocks as small as 2500 symbols.

Keywords: passive optical networks; nonlinear compensation; clustering

1. Introduction

The popularization of mobile multimedia applications and cloud computing, in combination with the emergence of the Internet of Things, is forcing telecommunications operators to increase the data capacity they can offer continuously. In this scenario, optical fibre infrastructures are progressively being extended, making them arriving closer to the end-user [1,2]. Among the different alternatives, passive optical networks (PONs) have been extensively employed due to their low cost [3–6]. Given the lack of amplification, however, fibre loss usually limits the bandwidth length product of these systems, especially when dispersion is compensated by predistortion techniques [7]. In order to extend the range of PON networks and, consequently, enable the resource concentration and cost reduction, two main approaches are commonly considered. On the one hand is the use of active splitting nodes that compensate the combined loss of fibre transmission and splitting [3,4]. This approach, nevertheless, makes the distribution network architecture more complicated and hinders its maintenance, particularly in long reach (LR) PONs where the splitting node may be far from urbanized areas. On the other hand, the advent of high speed digital signal processors (DSPs) has enabled the implementation of cost efficient digital coherent receivers [6], which dramatically changed the way about which high capacity links are thought. This inflexion point is not only due to the mixing gain they offer, but also because they permit simultaneous phase and amplitude modulation and open the possibility to compensate system impairments in a completely new and efficient way [8].

With linear impairments such as dispersion, phase noise, polarization fluctuation, and polarization-mode dispersion elegantly compensated in the baseband electrical domain, the interplay between nonlinear distortion and noise emerges as the main capacity limitation [9]. In single channel systems where the transmitted signal is broadcast to several users, the received power is reduced compared to wavelength division multiplexed (WDM) systems where demultiplexers are used and, consequently, the impact of the receiver noise is more critical. In regards to nonlinear distortion, the Kerr effect is widely claimed as the dominant nonlinear effect in digital coherent systems [10,11]. It is well known that the Kerr effect further leads to self-phase modulation (SPM), cross-phase modulation (XPM), and four wave mixing (FWM) processes [12]. In the case of multi-wavelength PON systems, it is expected that XPM and FWM will be the main nonlinear degradation mechanisms, but for those systems operating with a single channel, the nonlinear distortion is solely governed by SPM. SPM is particularly harmful in modulation formats with non-uniform amplitude, for instance 16 quadrature amplitude modulation (16-QAM) [13]. Since SPM causes a phase rotation that is proportional to the power of the symbol ($\phi_{NL} \approx -\gamma P L_{eff}$, where γ is the nonlinear coefficient, P is the symbol power, and L_{eff} is the effective fibre length), for moderate and elevated launch optical power levels, the constellation points with higher amplitude suffer a larger phase rotation than those with lower amplitude, leading to a characteristic spiral-like shape constellation [14]. This rotation, nevertheless, cannot be completely corrected by a simple nonlinear phase rotation because, even in a low dispersion regime, and the interplay of SPM and chromatic dispersion leads to more complex distortion. This complex distortion can be understood by noting that the pulse broadening caused by the chromatic distortion leads to word dependent behaviour through intersymbol interference (ISI). Even if ISI can be efficiently compensated in the baseband domain by DSP processing, the different superposed pulses are nonlinearly mixed through SPM. In addition, the interaction between the linear and nonlinear impairments varies as the signal propagates through the fibre. Thus, in the initial part of the fibre, the Kerr effect is significant, whereas the effect of the ISI created by the chromatic dispersion can be neglected. At the end of the fibre, on the other hand, the accumulated dispersion is high, leading to a significant ISI, but given the high transmission loss, the Kerr effect is reduced. Both scenarios can be modelled relatively easily, the initial part by a memoryless nonlinear phase rotation and the last part of the link by a linear time invariant system. The intermediate part of the link, however, should consider the interaction between the two effects, and therefore, it is complicated to model analytically.

Several nonlinear mitigation techniques have been proposed in recent years to overcome this issue. Optical techniques, for instance mid-span nonlinear compensation [15], lack flexibility and require a careful design of the distribution network, which is not possible in PON networks. In these networks, electrical compensation techniques are preferable for their higher flexibility and adaptability. Unfortunately, simple deterministic approaches based on non-uniform phase rotation cannot efficiently compensate signal distortion as they neglect the effect of chromatic dispersion and the subsequent ISI. More complex techniques, such as digital back-propagation (DBP) [16,17] and inverse Volterra series transfer function (IVSTF) based nonlinear equalization [18,19], were then studied to invert the dynamic nonlinear time invariant behaviour of the fibre link. These techniques are capable of mitigating the effect of the interplay between dispersion and nonlinear distortion, but suffer from a prohibitively high computational cost, making real-time operation if not unfeasible, at least extremely challenging and power consuming. In this scenario, machine learning emerges as a high potential set of tools to analyse and process complex systems where analytical modelling is unfeasible or the computational cost to solve it is excessively high [20]. Thus, several groups have proposed different machine learning based approaches to overcome the degrading effect of nonlinear distortion in fibre communication systems [21–24]. In [25–27], artificial neural networks were employed, whereas in [28] and in [29,30], support vector machines (SVMs) were proposed. These approaches present a good nonlinear compensation performance, but are all supervised and, consequently, require the transmission of a training sequence that reduces the effective data throughput. Unsupervised machine learning, for instance clustering, on the other hand, does not require any training sequence, but learns from the received dataset. Among the proposed clustering algorithms, k-means is by far the most popular due to its simplicity, its convergence speed, and robustness [31–34]. In k-means, however, each cluster is represented by a centroid, and the decision regions are limited by straight boundaries, which may not be optimal for constellations strongly affected by SPM. In order to find decision regions adapted to arbitrarily shaped clusters, other clustering algorithms have been proposed: In [35], clustering based on affinity propagation was reported. Density based spatial clustering of applications with noise (DBSCAN) has also been applied successfully to improve the performance of systems affected by the combined effect of noise and nonlinear distortion [36]. In [37], classification based on expectation maximization was successfully employed to combat the effect of nonlinear phase noise. These algorithms, however, suffer from heavy computational cost, and their performance strongly depends on the tuning parameters. In DBSCAN, for example, it is necessary to set the values of the parameters ϵ and k_{min} that correspond to the radius of the area and the minimum number of points in this area, respectively.

In this paper, we present a novel clustering algorithm denominated histogram based clustering (HBC) that partially mitigates the effect of nonlinear phase noise caused by SPM. The proposed approach is a density based clustering algorithm that assigns to a received symbol the class of the closest high point-density region. This is different from other clustering algorithms as k-means, which neglects any density information, or expectation maximization, which estimates the point distribution as a mixture of Gaussian distributions. Compared to the main density based clustering algorithm, DBSCAN, it does not require the setting of ϵ and k_{min}; it is able to find the best point density value automatically to have the desired number of clusters. In addition, the adopted solution is not iterative and leads to deterministic complexity. The rest of the paper is organized as follows: Section 2 explains the proposed clustering algorithm. In Section 3, the simulation setup is described, while in Section 4, the results of applying HBC to an LR-PON network are presented and discussed, paying attention not only to its performance, but also to the required block size. Section 4, finally, concludes the paper.

2. Histogram Based Clustering

Figure 1 shows the flow diagram of the proposed HBC algorithm. For the sake of illustration, we employed 10,000 16-QAM symbols that were obtained by applying an amplitude dependent nonlinear phase rotation ($\Delta\phi[n] = 3A^2[n]$, where $A[n]$ is the amplitude of the symbol) and including complex additive noise with a signal-to-noise ratio (SNR) of 20 dB. It is worth noting that this simple model was employed only for demonstration purposes because by neglecting the intersymbol interference, the model did not accurately represent the system. In the next section, we will describe the model used to consider simultaneously nonlinear and linear effects that result in more complex symbol distributions. The resultant constellation is shown in Figure 1a. The histogram of the unlabelled symbols was calculated, in this case, using 40 bins in the in-phase and quadrature directions, resulting in the contour plot presented in Figure 1b, where the 16 clusters can be clearly identified.

Once the histogram was calculated, the lowest value contour line led to the desired number of clusters, 16 in the case of 16-QAM. The value of this contour line represents the optimal point density. A lower point density does not allow the correct recovery of all the clusters, while a larger value results in a too conservative criterion that leads to poorer performance, as the decision regions are not well matched to the data. This optimal density searching mechanism is, indeed, one of the main advantages over other density based clustering algorithms, such as DBSCAN. In fact, DBSCAN is intended for data clustering where the number of clusters is a priori unknown and the minimum density of clusters is fixed through the values of ϵ and k_{min}. This was not our case, where the number of clusters was known and the minimum density was dependent on the distortion and, therefore, also on the launch power level. The determined cluster borders are superimposed on the histogram in Figure 1c, showing that, especially the constellation points with stronger distortion, those in the periphery, the boundaries of the different clusters were closer to each other. After finding the cluster boundaries, each cluster was identified by the points that formed its boundary instead of a centroid as in k-means. Figure 1d shows the boundaries of the different clusters on top of the received constellation, where it can be clearly seen that the obtained cluster boundaries encompassed most of the points of their respective clusters. Once the boundary points for each cluster were found, the distance from each received symbol to them was calculated, and the class of the boundary point with the shortest distance was assigned to the symbol. The classified constellation is shown in Figure 1e.

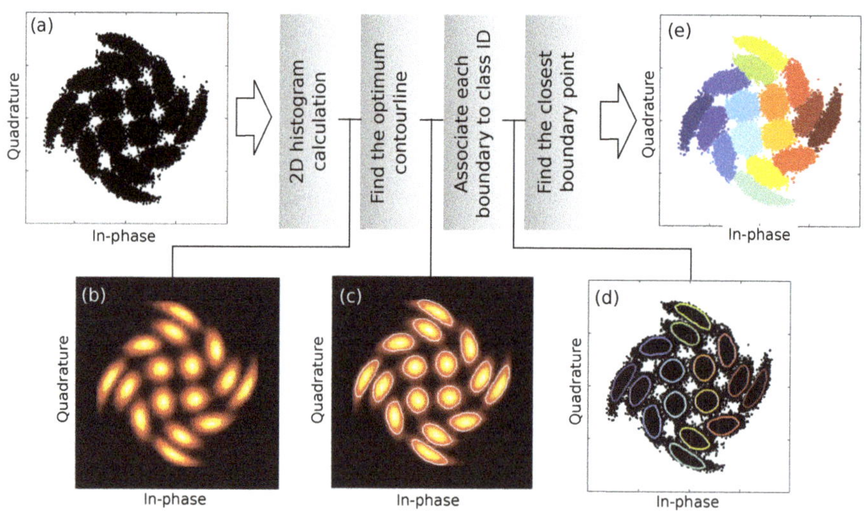

Figure 1. Flow diagram of the proposed clustering algorithm. (**a**) Distorted input 16-QAM constellation. (**b**) Calculated 2D histogram. (**c**) Optimum boundaries superimposed on the 2D histogram. (**d**) Boundaries for each cluster on top of the received distorted constellation. (**e**) Classified constellation.

In a systematic way, the proposed HBC algorithm consisted of the following four steps:

1. Calculation of the 2D histogram of the in-phase and quadrature components of the N_S received distorted symbols.
2. Find the lowest contour line in the histogram that results in M isolated islands, M being the number of clusters to be identified.
3. Assign a class ID to the values of the boundary for each island.
4. For each received symbol, find the closest boundary point and associate it with its class ID.

3. Simulation Setup

The performance of the proposed HBC algorithm was assessed employing the simulation setup presented in Figure 2, where the electrical modulation and demodulation tasks were implemented in MATLAB, while the conversion between the electrical and optical domain, as well as the transmission through the passive distribution network were carried out in VPI Transmission Maker.

On the transmitter side, a 1 mW power continuous wave (CW) laser diode (LD_1) operating at 1550 nm was externally modulated using a dual parallel Mach–Zehnder modulator (DP-MZM) driven by the in-phase and quadrature components of a 56 Gbps 16-QAM signal filtered by a fourth order Bessel filter with a bandwidth of 10.5 GHz. The modulated optical signal was then amplified by an erbium doped fibre amplifier (EDFA), and a variable optical attenuator was used to vary the launch optical power between 2 and 12 mW. Since the output power of the EDFA was fixed, the noise amplified spontaneous emission (ASE) noise added by the amplifier remained constant. Furthermore, given the relatively low gain of the amplifier, the signal-ASE beating at the output of the receiver was negligible compared to the receiver noise.

The distribution network was simulated using a first span of standard single mode fibre (SSMF) that had a length of 80 km, a one-to-64 splitter (emulated by an 18 dB attenuator), and a second SSMF span with a 20 km length.

The coherent receiver was formed by an optical front-end where the state of polarization of the received signal was first controlled using a dynamic polarization tracker (DPT), which made the signal polarization match that of the local oscillator. The signal was then combined in a 90°-hybrid network with a 1 mW power CW laser (LD_2). The combined signals were photodetected and filtered before being differentially amplified. Analogue-to-digital conversion (ADC) was emulated by downsampling the signal to four samples per symbol, after which frequency domain chromatic dispersion (CD) compensation was performed. Afterwards, the synchronization of the signal was performed by the cross-correlation maximization method using an alternated synchronization sequence of 64 symbols. In order to reduce the overhead, this synchronization sequence was also employed for amplitude scaling and initial phase synchronization. After CD mitigation, synchronization, and scaling, the signal underwent a second downsampling process in order to get a single sample per symbol. Phase noise correction was performed by blind phase search operating on 32 symbol blocks [38]. In our simulations, we did not consider polarization mode dispersion (PMD) because, in contrast to CD, it does not significantly interact with nonlinear distortion and could be satisfactorily compensated in variable envelope modulations using, for example, the multiple modulus algorithm (MMA) [39]. The distorted constellations were then processed using the proposed HBC algorithm. For comparison purposes, we also present results considering maximum likelihood, as well as k-means, which are considered as benchmarks for linear and clustered detection, respectively.

Regarding the performance metric, we adopted the bit error rate (BER), which taking into account the lack of Gaussianity of the constellation point distribution, had to be estimated by error counting. In addition, we calculated the equivalent Q-factor derived from the BER according to: $Q = \sqrt{2} \cdot erfc^{-1}(2BER)$, where $erfc^{-1}$ denotes the inverse complementary error function.

Table 1 lists the most important simulation parameters.

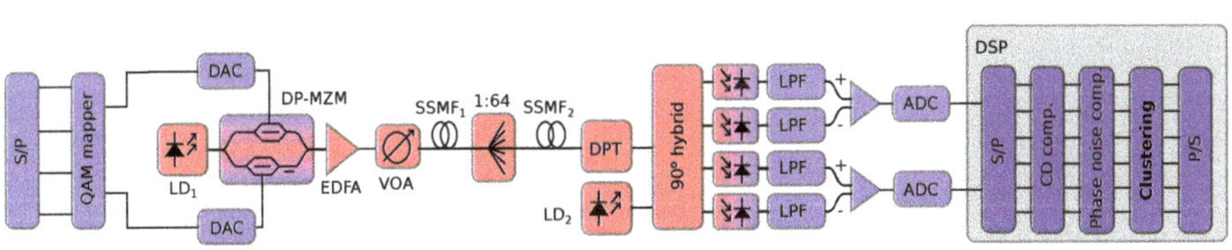

Figure 2. Block diagram of the simulated coherent LR-PON system. S/P: serial-to-parallel conversion. DAC: digital-to-analogue converter. LD: laser diode. DP-MZM: dual parallel Mach–Zehnder modulator. EDFA: erbium doped fibre amplifier. VOA: variable optical attenuator. SSMF: standard single mode fibre. DPT: Dynamic polarization tracker. LPF: low pass filter. ADC: analogue-to-digital converter.

Table 1. Parameters used in the simulation.

System Parameters			
Laser linewidth	0.5 MHz	Fibre lengths (L_1,L_2)	80 km, 0–20 km
Laser power	1 mW	Fibre attenuation	0.2 dB
MZM insertion loss	6 dB	Fibre chromatic dispersion	16 ps/nm/km
Amplifier gain	20 dB	Fibre PMD	3.16 fs/$\sqrt{\text{km}}$
Amplifier noise figure	4 dB	Nonlinear coefficient (γ)	$1.3 \cdot W^{-1} \cdot km^{-1}$
Attenuator	20 dB	Fibre effective area	80 μm^2
PD thermal noise density	10 pA/$\sqrt{\text{Hz}}$	Electrical filter bandwidth	10.5 GHz
PD responsivity	1 W/A	Electrical RX filter order	4
Signal parameters			
Modulation format	16-QAM	No. of synchronization symbols	64
Electrical TX filter	4th-order Bessel	Bit rate	56 Gbps
Simulation parameters			
Number of simulated symbols	16,384	Sampling rate	8.96×10^{11} s^{-1}

4. Results and Discussion

4.1. Performance Analysis

In order to analyse the performance of the proposed HBC algorithm, in Figure 3a,b, we show the BER and the corresponding Q-factor at launch optical powers ranging from 2 to 12 mW for the three different approaches: maximum likelihood detection, clustering using k-means, and the novel HBC algorithm. At low power levels, the performance of all three techniques improved (BER reduced and Q-factor increased) as the launch optical power was increased, which made sense since the additive noise of the photodetectors was dominant.

For high power levels, on the other hand, Kerr induced SPM was the main physical impairment, and consequently, increasing power led to higher BER and a lower Q-factor. Comparing the performance of maximum likelihood with k-means and HBC, it was clear that all of them converged for a low power level, while the latter two presented improved performances for high power levels. This was an indicator that k-means and HBC were indeed compensating nonlinear distortion and not any linear impairment such as residual phase noise or chromatic dispersion. These two clustering algorithms, however, showed different performances, for both medium and high power levels. It can be clearly observed that HBC outperformed k-means for launch optical powers above 5 mW, revealing that HBC could mitigate nonlinear distortion more efficiently than the traditional k-means clustering. As a result, the best achievable BER was reduced from 1.1×10^{-3} when using maximum likelihood and 0.8×10^{-3} when employing k-means to 0.6×10^{-3} in the case of HBC (orange polygon of Figure 3a). Regarding the Q-factor, numerical results showed an improvement of 0.53 dB with respect to the optimum performance of maximum likelihood and 0.23 dB when contrasted with the optimum of k-means (orange polygon of Figure 3b). The Q-factor enhancement was higher if, instead of comparing optimum performances, we looked at a fixed power level in the nonlinear regime. Thus, for 10 mW, HBC outperformed k-means by 0.51 dB and maximum likelihood by 1.22 dB (purple polygon of Figure 3b). Additionally, the optimum launch optical power where the trade-off between noise and SPM was held shifted towards a higher power level, from 5 mW in maximum likelihood to 6 and 7 mW for k-means and HBC, respectively.

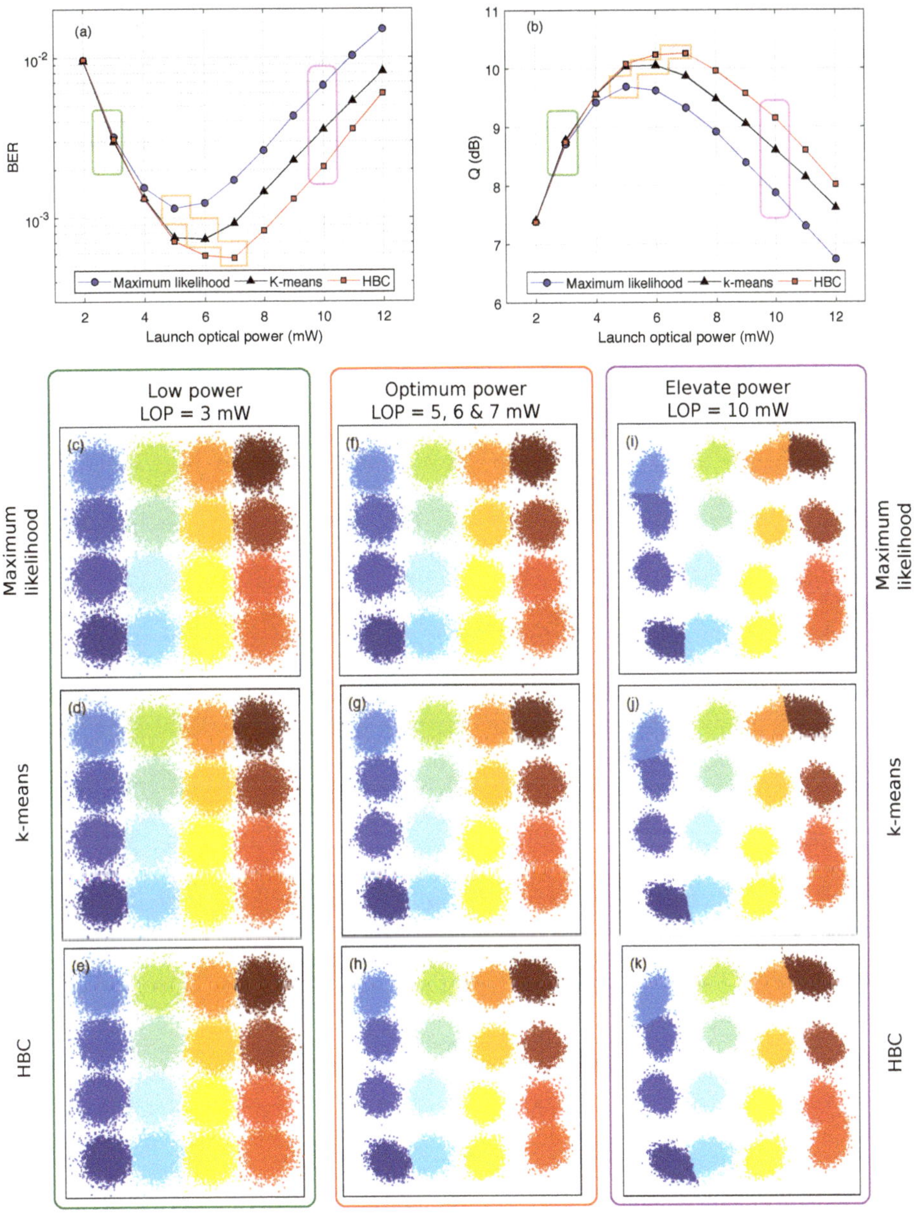

Figure 3. Performance of the proposed. HBC compared to that of maximum likelihood and k-means in terms of (**a**) BER and (**b**) effective Q-factor derived from BER. (**c**,**d**,**e**) Classified constellations using maximum likelihood, k-means, and HBC, respectively, for a launch optical power of 3 mW. Classified constellations at optimum launch optical powers for the different detection schemes: (**f**) 5 mW for maximum likelihood, (**g**) 6 mW for k-means, and (**h**) 7 mW for HBC. (**i**,**j**,**k**) Classified constellations at elevated launched optical power (10 mW) for maximum likelihood detection, k-means, and HBC, respectively.

A better understanding of the performance enhancement can be achieved by analysing the classified constellations (after undergoing maximum likelihood, k-means, or HBC) for different power levels. In particular, Figure 3c–e represents the classified constellations when using maximum likelihood, k-means, or HBC, respectively, for a launch optical power of 3 mW (green rectangle). Figure 3f–h shows the constellation for the optimum power levels, that is 5 mW for maximum likelihood, 6 mW for k-means, and 7 mW for HBC (orange polygon). The constellation degradation and classification for a relatively large power level, 10 mW, can be observed in Figure 3i–k (purple rectangle). Looking at the point dispersion for different power levels, e.g., 3, 5, and 10 mW, shown in the upper row, it can be seen that whereas for a low power level, the constellation shape remained, as the launch power increased, SPM distorted the constellation leading to non-rectangular constellations. The inner symbols seemed to be rotated clockwise, in contrast to the symbols in the periphery that were rotated counter-clockwise. In fact, all the symbols were rotated counter-clockwise by the Kerr effect, but the phase noise compensation stage in the DSP inverted the average rotation, resulting in some points (those with smaller rotation corresponding to lower power) being rotated in the opposite direction. Another feature to be noted is that the noise variance at low power levels looked higher than for moderate power levels, but since the main noise mechanisms were the thermal and shot noises of the photodetectors, the noise level was virtually the same for the three power levels. This is typical in power limited coherent systems because the optical power arriving at the photodetectors is mainly that of the local oscillator laser. The apparently lower noise was then a consequence of the higher signal power and of the power normalization performed before clustering was carried out. Comparing the performance of maximum likelihood, k-means, and HBC, the reader can observe that for low power levels, the classifications obtained by the three methods were essentially identical, which agreed with the fact that same BER and Q-factor values were yielded. This made sense because, in the absence of non linear distortion, maximum likelihood was the optimum detection scheme [40]. As the power level increased, so did the SNR, but SPM led to the aforementioned symbol rotation. Is in this case, the rectangular decision regions of maximum likelihood were not optimum any more, and clustering with non-rectangular boundaries fit the distorted data better. At even higher power levels, SPM led to more complicated cluster sizes where decision regions with linear boundaries, as those obtained using k-means, may result in sub-optimal classification.

The differences between the resultant decision regions using maximum likelihood, k-means, and HBC can be better observed in Figure 4, where to make the contrast more clear, data for a launch optical power of 10 mW were employed. First of all, we show the histogram of the received constellation in Figure 4a to demonstrate how the distortion especially affected the symbols with higher amplitude. A detailed view of two of the constellation points that were critically affected by SPM (identified by a white rectangle) can be seen under the constellation plot. As can be observed, the two clusters presented a complex shape, and therefore, an intricate border was necessary to classify them. Figure 4b shows the constellation of the received data superimposed on the decision regions calculated using maximum likelihood. As expected, the rectangular grid led to multiple constellation point to invade adjacent regions even for low power symbols. The zoom-in figure clearly reveals that the decision boundary was not optimal for the symbol distribution. When employing k-means clustering, more sophisticated regions were calculated, as shown in Figure 4c, and the low power constellation points were then correctly classified. For symbols in the constellation periphery, the straight boundaries of k-means, however, were not optimal. This point can be appreciated in the zoom-in. Finally, if we applied HBC, we obtained the decision regions represented in Figure 4d. At first glance, the decision regions were similar to those using k-means. For low power symbols, where SPM did not significantly affect the shape, but caused only a rotation (this can be corroborated in the histogram of Figure 4a), the boundaries were still straight lines. For higher power symbols, in contrast, the boundaries found by HBC were not straight lines any more. We can see a clear example of the curved boundaries in the detailed view, appreciating that the curved line matched better the cluster boundary expected from the histogram.

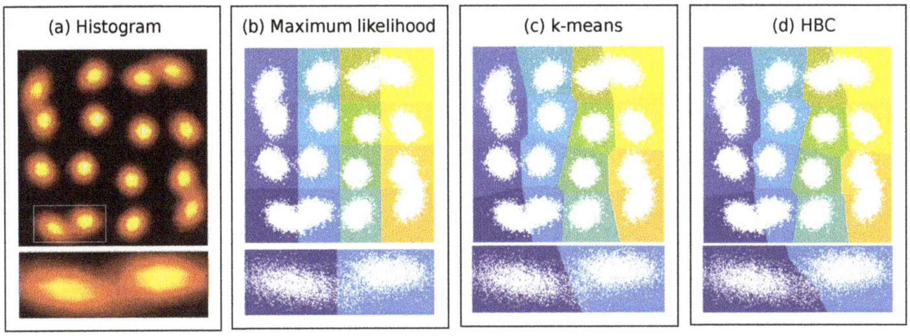

Figure 4. Analysis of the decision regions (data are for a launch optical power of 10 mW). (**a**) Histogram of the received constellation. (**b**) Rectangular shaped decision regions obtained using maximum likelihood. (**c**) Linear decision regions after k-means clustering and (**d**) after HBC. In (**b–d**), the received constellations are superimposed on the decision regions. In addition, for all cases, we included a detailed section of the lower left constellation, corresponding to the white rectangle in (**a**).

4.2. Block Size and Complexity Analysis

The proposed HBC algorithm, as already mentioned, can be regarded as a non-iterative density based clustering algorithm. The fact of not requiring iteration was of particular importance for real-time operation since, in principle, it implies a fixed processing time. The performance of the algorithm, however, depended on the block size employed to estimate the histogram. To evaluate the minimum block size requirement, Figure 5 shows the BER in terms of the block size. As can be seen after a short erratic initial stage, the BER decreased, getting a relatively stable value for block sizes longer than 13,000. The evaluation of the performance for small block sizes was especially difficult because of the high variance consequence of the stochastic nature of the method. We observed that the variance of the BER reduced as the block size increased, as was expected for unbiased statistics. For this reason, to achieve a trade-off between accuracy and processing time, the data shown in Figure 5 were obtained by averaging results for a variable number of runs, up to 100 for 1000 symbol block size and five for 16,000 symbol blocks. In fact, we perceived that when applying HBC for short block sizes, for certain sets of data, the algorithm did not converge to a reasonable partition. In order to quantify this effect, for each block size, we counted the number of runs that failed, and we calculated the efficiency as the percentage of runs that led to an acceptable classification. This region of forbidden block size that, according to Figure 5, spanned up to 2250, should then be avoided to get an acceptable performance. In fact, for applications requiring low latency, we can choose the minimum size of 2250, whereas when latency is not so critical, a higher number of symbols can be employed.

Another important point to discuss is the complexity of the proposed algorithm. In order to evaluate it, we can split the algorithm into two main steps: on the one hand, a first stage when the 2D histogram was built and the high density points were found and, on the other hand, the stage when each point was associated with a certain cluster. The histogram can be built in different ways, so it was expected to be machine dependent. A possible solution was, for instance, to find the indexes of the bins for each symbol and, then, update the value of the corresponding bin. That is, assuming that we had N_b bins and that the maximum and the minimum of the histogram were M and m, the indexes corresponding to the kth complex symbol $s[k] = s_i[k] + j \cdot s_q[k]$ were:

$$ind_i[k] = \frac{N_b + 1}{M - m} \cdot (s_i[k] - m) \text{ and } ind_q[k] = \frac{N_b + 1}{M - m} \cdot (s_q[k] - m). \quad (1)$$

Once the indexes were found, the count of the bin indicated by ind_i and ind_q, n^{k-1}_{count} was updated. Hence:

$$n^{k}_{count}(ind_i, ind_q) = n^{k-1}_{count}(ind_i, ind_q) + 1. \quad (2)$$

Therefore, in this first stage, the processing of each symbol required five floating-point operations (two to calculate ind_i, two to calculate ind_q, and another one to update the bin count). In order to build the histogram of a block of N_{sym} symbols, the total number of FLOPs was then $5 \cdot N_{sym}$, and in conclusion, its complexity was $\mathcal{O}(N_{sym})$. The finding of the high density points required the sorting of the values of all bins, that is the sorting of N_b^2 points. Sorting algorithms, such as block based or binary tree, present a complexity of $\mathcal{O}(N_b^2 \log N_b^2)$. Therefore, the complexity will depend on the number of symbols and employed bins. In our case, we employed 100,000 symbols and 40 bins, and as a consequence, the histogram building process was the dominant term. The complexity of the second stage, that is finding the closest high density point to a given symbol, depends on the number of high density points. Furthermore, this number will vary depending on the shape of the clusters. In particular, the noisier the clusters are, the larger the areas of relative high density and the number of points with high density are. If we assume that we have a set of S_{hd} of N_{hd} points of high density, then for each symbol, we need to calculate N_{hd} distances (indeed, it is possible to calculate the square of the distance):

$$D = d^2[k] = (s_i[k] - u_i[m])^2 + (s_q[k] - u_q[m])^2, \text{ where } u[m] = u_i[m] + j \cdot u_q[m] \in S_{hd}. \quad (3)$$

This distance required five real valued operations, as we needed to calculate 2 subtractions, 2 multiplications, and 1 addition. The complexity of this stage was then $\mathcal{O}(N_s \cdot N_{hd})$, which was higher than that of k-means, where the number of distances to be calculated corresponded to the number of clusters. However, it should be noted that this comparison only considered the distance calculation and not the number of operation to find the centroids. The complexity of HBC was also higher than that of simple nonlinear rotation, which was as small as two real valued operations.

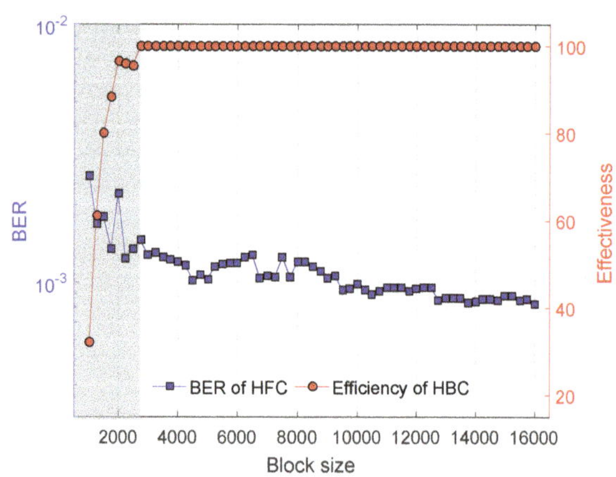

Figure 5. BER in terms of the processed block size alongside with the efficiency of HBC for a power level of 7 mW at which optimum performance is achieved for HBC. For comparison purposes, the BER obtained using maximum likelihood is also included.

4.3. Discussion

In this paper, we tested the proposed HBC algorithm in a single channel coherent LR-PON, where the transmitted signal was distorted by the combination of SPM and the noise added by the receiver. As can be appreciated in the constellations of Figure 3c–k, the obtained symbols showed the characteristic spiral-like constellations that were prone to be clustered. The high distribution losses, which included the fibre and the splitting losses, however, made the constellations to have noise and the cluster borders to appear blurred. This was not the case of WDM systems, where the lack of splitters leads to higher received power and the subsequent reduction of the receiver noise impact. Indeed, for the configuration considered in this work, the launch optical power had to be increased well above the optimum power in order for the XPM and FWM to cause a distortion comparable to the excess splitting loss. In regards to dual polarization operation, the proposed algorithm could be modified to account for both polarizations simultaneously. This dimensionality increase, however, would lead to more complicated processing.

The proposed algorithm aimed to compensate the distortion in simple LR-PON systems without requiring high computational cost and a training sequence. In this sense, HBC can be considered as a trade-off between performance and complexity. Indeed, we can note that HBC outperformed the nonlinear phase rotation at the cost of higher complexity and latency. On the other hand, HBC presented a slightly worse tolerance to nonlinear distortion than other more sophisticated algorithms (for example, 2.5 dB of HBC vs. 3 dB of EM [37]), but without requiring initialization and the iterative process.

Another point to be considered is the employed filters, in our simulations fourth order Bessel filters. These filters emulate the bandwidth limitation of both the transmitter electronics and PD response and remove part of the out-of-band noise. It is envisaged that the adoption of more sophisticated filters, Nyquist filtering in particular, could improve the performance, as they increase the SNR, thus making the clusters easier to discriminate.

5. Conclusions

In this paper, we proposed a novel clustering algorithm based on histograms, which we denominated histogram based clustering. The algorithm successfully compensated the distortion caused by Kerr mediated SPM in coherent LR-PONs with a transmission distance of 100 km and a splitting ratio of 64. The numerical results obtained using VPI Transmission Maker-MATLAB co-simulation showed that HBC improved the Q-factor with respect to maximum likelihood and k-means clustering by 0.53 dB and 0.23 dB, respectively. We also showed that the proposed algorithm could operate on blocks of 2500 symbols, but that optimum performance was obtained for blocks of 12,000 symbols.

Author Contributions: Conceptualization, I.A., E.G., and M.L.F.A.; methodology, I.A. and E.G.; software, I.A. and G.d.O.; validation, I.A. and M.L.F.A; formal analysis, J.W., J.L.P., J.D.M., and E.A.M.F.; resources, I.A., E.G., L.B., and M.L.F.A.; writing–original draft preparation, I.A., J.L.P. and E.A.M.F.; writing–review and editing, I.A., E.G., G.d.O., J. W., J.L.P., J.D.M., E.A.M.F., L.B., and M.L.F.A; visualization, I.A., E.G., G.d.O., J. W., J.L.P., J.D.M., E.A.M.F., L.B., and M.L.F.A; supervision, I.A. and M.L.F.A.; project administration, I.A.; funding acquisition, I.A., E.G., L.B., and M.L.F.A. All authors have read and agreed to the published version of the manuscript.

Funding: This research was funded by the National Council for Scientific and Technological Development (CNPQ) of Brazil, the EU Horizon 2020 Research and Innovation Programme through the Marie Sklodowska-Curie under Grant 713567, the Science Foundation of Ireland, and the European Regional Development Fund under Grant 13/RC/2077.

Conflicts of Interest: The authors declare no conflict of interest.

Abbreviations

The following abbreviations are used in this manuscript:

PON	Passive optical network
LR	Long reach
DSP	Digital signal processor
SPM	Self-phase modulation
XPM	Cross-phase modulation
FWM	Four wave mixing
QAM	Quadrature amplitude modulation
DBP	Digital back-propagation
IVSTF	Inverse Volterra series transfer function

SVM	Support vector machine
DBSCAN	Density based spatial clustering of applications with noise
HBC	Histogram based clustering
LD	Laser diode
CW	Continuous-wave
DP-MZM	Dual parallel Mach–Zehnder modulator
EDFA	Erbium doped fibre amplifier
ASE	Amplified spontaneous emission
DAC	Digital to analogue converter
SSMF	Standard single mode fibre
DPT	Dynamic polarization tracker
ADC	Analogue-to-digital converter
BER	Bit error rate
SNR	Signal-to-noise ratio

References

1. Nesset, D. PON roadmap. *IEEE/OSA J. Opt. Commun. Netw.* **2017**, *9*, A71–A76. [CrossRef]
2. Effenberger, F.J. Industrial trends and roadmap of access. *J. Lightwave Technol.* **2017**, *35*, 1142–1146. [CrossRef]
3. Hsu, D.Z.; Wei, C.C.; Chen, H.Y.; Chen, J.; Yuang, M.C.; Lin, S.H.; Li, W.Y. 21 Gb/s after 100 km OFDM long reach PON transmission using a cost-effective electro-absorption modulator. *Opt. Express* **2010**, *18*, 27758–27763. [CrossRef] [PubMed]
4. Chen, M.; He, J.; Chen, L. Real-time optical OFDM long reach PON system over 100 km SSMF using a directly modulated DFB laser. *J. Opt. Commun. Netw.* **2014**, *6*, 18–25. [CrossRef]
5. Lavery, D.; Ionescu, M.; Makovejs, S.; Torrengo, E.; Savory, S.J. A long reach ultra-dense 10 Gbit/s WDM-PON using a digital coherent receiver. *Opt. Express* **2010**, *18*, 25855–25860. [CrossRef] [PubMed]
6. Lavery, D.; Maher, R.; Millar, D.S.; Thomsen, B.C.; Bayvel, P.; Savory, S.J. Digital coherent receivers for long reach optical access networks. *J. Lightwave Technol.* **2013**, *31*, 609–620. [CrossRef]
7. Liu, X.; Effenberger, F. ONU-dependent dispersion pre-compensation at OLT for high speed wide-coverage PON. In Proceedings of the 2015 IEEE Global Communications Conference (GLOBECOM), San Diego, CA, USA, 6–10 December 2015; pp. 1–5.
8. Kikuchi, K. Fundamentals of coherent optical fibre communications. *J. Lightwave Technol.* **2016**, *34*, 157–179. [CrossRef]
9. Kikuchi, K.; Tsukamoto, S. Evaluation of sensitivity of the digital coherent receiver. *J. Lightwave Technol.* **2008**, *26*, 1817–1822. [CrossRef]
10. Ip, E.M.; Kahn, J.M. Fibre impairment compensation using coherent detection and digital signal processing. *J. Lightwave Technol.* **2009**, *28*, 502–519. [CrossRef]
11. Vacondio, F.; Rival, O.; Simonneau, C.; Grellier, E.; Bononi, A.; Lorcy, L.; Antona, J.C.; Bigo, S. On nonlinear distortions of highly dispersive optical coherent systems. *Opt. Express* **2012**, *20*, 1022–1032. [CrossRef]
12. Agrawal, G.P. Nonlinear fibre optics. In *Nonlinear Science at the Dawn of the 21st Century*; Springer: Berlin, Germany, 2000; pp. 195–211.
13. Gordon, J.P.; Mollenauer, L.F. Phase noise in photonic communications systems using linear amplifiers. *Opt. Lett.* **1990**, *15*, 1351–1353. [CrossRef] [PubMed]
14. Lau, A.P.T.; Kahn, J.M. Signal design and detection in presence of nonlinear phase noise. *J. Lightwave Technol.* **2007**, *25*, 3008–3016. [CrossRef]
15. Da Ros, F.; Sackey, I.; Elschner, R.; Richter, T.; Meuer, C.; Nölle, M.; Jazayerifar, M.; Petermann, K.; Peucheret, C.; Schubert, C. Kerr nonlinearity compensation in a 5× 28-GBd PDM 16-QAM WDM system using fibre based optical phase conjugation. In Proceedings of the 2014 The European Conference on Optical Communication (ECOC), Cannes, France, 21–25 September 2014; pp. 1–3.
16. Ip, E.; Kahn, J.M. Compensation of dispersion and nonlinear impairments using digital backpropagation. *J. Lightwave Technol.* **2008**, *26*, 3416–3425. [CrossRef]

17. Millar, D.S.; Makovejs, S.; Behrens, C.; Hellerbrand, S.; Killey, R.I.; Bayvel, P.; Savory, S.J. Mitigation of fibre nonlinearity using a digital coherent receiver. *IEEE J. Sel. Top. Quantum Electron.* **2010**, *16*, 1217–1226. [CrossRef]
18. Pan, J.; Cheng, C.H. Nonlinear electrical predistortion and equalization for the coherent optical communication system. *J. Lightwave Technol.* **2011**, *29*, 2785–2789. [CrossRef]
19. Giacoumidis, E.; Aldaya, I.; Jarajreh, M.A.; Tsokanos, A.; Le, S.T.; Farjady, F.; Jaouën, Y.; Ellis, A.D.; Doran, N.J. Volterra based reconfigurable nonlinear equalizer for coherent OFDM. *IEEE Photonics Technol. Lett.* **2014**, *26*, 1383–1386. [CrossRef]
20. Alpaydin, E. *Introduction to Machine Learning*; MIT Press: Cambridge, MA, USA, 2009.
21. Zibar, D.; Piels, M.; Jones, R.; Schäeffer, C.G. Machine learning techniques in optical communication. *J. Lightwave Technol.* **2015**, *34*, 1442–1452. [CrossRef]
22. Giacoumidis, E.; Lin, Y.; Wei, J.; Aldaya, I.; Tsokanos, A.; Barry, L. Harnessing machine learning for fibre-induced nonlinearity mitigation in long-haul coherent optical OFDM. *Future Internet* **2019**, *11*, 2. [CrossRef]
23. Khan, F.N.; Lu, C.; Lau, A.P.T. Machine learning methods for optical communication systems. In *Signal Processing in Photonic Communications*; Optical Society of America: Washington, DC, USA, 2017; p. SpW2F-3.
24. Mata, J.; De Miguel, I.; Duran, R.J.; Merayo, N.; Singh, S.K.; Jukan, A.; Chamania, M. Artificial intelligence (AI) methods in optical networks: A comprehensive survey. *Opt. Switch. Netw.* **2018**, *28*, 43–57. [CrossRef]
25. Ahmad, S.T.; Kumar, K.P. Radial basis function neural network nonlinear equalizer for 16-QAM coherent optical OFDM. *IEEE Photonics Technol. Lett.* **2016**, *28*, 2507–2510. [CrossRef]
26. Chen, E.; Tao, R.; Zhao, X. Channel equalization for OFDM system based on the BP neural network. In Proceedings of the 2006 8th international Conference on Signal Processing, Guilin, China, 16–20 November 2006; Volume 3.
27. Jarajreh, M.A.; Giacoumidis, E.; Aldaya, I.; Le, S.T.; Tsokanos, A.; Ghassemlooy, Z.; Doran, N.J. Artificial neural network nonlinear equalizer for coherent optical OFDM. *IEEE Photonics Technol. Lett.* **2014**, *27*, 387–390. [CrossRef]
28. Nguyen, T.; Mhatli, S.; Giacoumidis, E.; Van Compernolle, L.; Wuilpart, M.; Mégret, P. Fibre nonlinearity equalizer based on support vector classification for coherent optical OFDM. *IEEE Photonics J.* **2016**, *8*, 1–9. [CrossRef]
29. Wang, D.; Zhang, M.; Li, Z.; Cui, Y.; Liu, J.; Yang, Y.; Wang, H. Nonlinear decision boundary created by a machine learning based classifier to mitigate nonlinear phase noise. In Proceedings of the 2015 European Conference on Optical Communication (ECOC), Valencia, Spain, 27 September–1 October 2015; pp. 1–3.
30. Giacoumidis, E.; Mhatli, S.; Nguyen, T.; Le, S.; Aldaya, I.; McCarthy, M.; Eggleton, B. Kerr-induced nonlinearity reduction in coherent optical OFDM by low complexity support vector machine regression based equalization. In Proceedings of the 2016 Optical Fibre Communications Conference and Exhibition (OFC), Anaheim, CA, USA, 20 – 24 March 2016; pp. 1–3.
31. Zhang, J.; Chen, W.; Gao, M.; Shen, G. K-means-clustering based fibre nonlinearity equalization techniques for 64-QAM coherent optical communication system. *Opt. Express* **2017**, *25*, 27570–27580. [CrossRef] [PubMed]
32. Li, M.; Yu, S.; Yang, J.; Chen, Z.; Han, Y.; Gu, W. Nonparameter nonlinear phase noise mitigation by using M-ary support vector machine for coherent optical systems. *IEEE Photonics J.* **2013**, *5*, 7800312. [CrossRef]
33. Boada, R.; Borkowski, R.; Monroy, I.T. Clustering algorithms for Stokes space modulation format recognition. *Opt. Express* **2015**, *23*, 15521–15531. [CrossRef]
34. Pakala, L.; Schmauss, B. Non-linear mitigation using carrier phase estimation and K-means clustering. In Proceedings of the Photonic Networks; 16. ITG Symposium, Leipzig, Germany, 7 – 8 May 2015; pp. 1–3.
35. Giacoumidis, E.; Aldaya, I.; Wei, J.; Sanchez, C.; Mrabet, H.; Barry, L.P. Affinity propagation clustering for blind nonlinearity compensation in coherent optical OFDM. In Proceedings of *CLEO: Science and Innovations*; San Jose, CA, USA, 15 – 17 May 2018 p. STh1C-5.
36. Giacoumidis, E.; Lin, Y.; Barry, L.P. Fibre Nonlinear Compensation Using Machine Learning Clustering. In Proceedings of the 56th ICREIT conference, London, UK, 18–19 January 2019, pp. 1–4.

37. Zibar, D.; Winther, O.; Franceschi, N.; Borkowski, R.; Caballero, A.; Arlunno, V.; Schmidt, M.N.; Gonzales, N.G.; Mao, B.; Ye, Y.; et al. Nonlinear impairment compensation using expectation maximization for dispersion managed and unmanaged PDM 16-QAM transmission. *Opt. Express* **2012**, *20*, B181–B196. [CrossRef]
38. Pfau, T.; Hoffmann, S.; Noé, R. Hardware-efficient coherent digital receiver concept with feedforward carrier recovery for *M*-QAM constellations. *J. Lightwave Technol.* **2009**, *27*, 989–999. [CrossRef]
39. Yang, J.; Werner, J.J.; Dumont, G.A. The multimodulus blind equalization and its generalized algorithms. *IEEE J. Sel. Areas Commun.* **2002**, *20*, 997–1015. [CrossRef]
40. Lathi, B.P. *Modern Digital and Analogue Communication Systems*; Oxford University Press, Inc.: Oxford, UK, 1998.

applied sciences

MDPI

Article

A Blind Nonlinearity Compensator Using DBSCAN Clustering for Coherent Optical Transmission Systems

Elias Giacoumidis [1,*], Yi Lin [1], Mutsam Jarajreh [2], Sean O'Duill [1], Kevin McGuinness [3], Paul F. Whelan [4] and Liam P. Barry [1]

1 Radio and Optical Laboratory, School of Electronic Engineering, Dublin City University, Glasnevin 9, Dublin, Ireland; yi.lin6@mail.dcu.ie (Y.L.); sean.oduill@dcu.ie (S.O.); liam.barry@dcu.ie (L.P.B.)
2 Computer Engineering Department, Fahad Bin Sultan University, Tabuk 47721, Saudi Arabia; mutsam.jarajreh@gmail.com
3 Insight Centre for Data Analytics, School of Electronic Engineering, Dublin City University, Dublin 9, Ireland; kevin.mcguinness@dcu.ie
4 Vision Systems Group, School of Electronic Engineering, Dublin City University, Dublin 9, Ireland; Paul.Whelan@dcu.ie
* Correspondence: elias.giacoumidis@dcu.ie; Tel.: +353-1-700-8598

Received: 29 August 2019; Accepted: 7 October 2019; Published: 17 October 2019

Abstract: Coherent fiber-optic communication systems are limited by the Kerr-induced nonlinearity. Benchmark optical and digital nonlinearity compensation techniques are typically complex and tackle deterministic-induced nonlinearities. However, these techniques ignore the impact of stochastic nonlinear distortions in the network, such as the interaction of fiber nonlinearity with amplified spontaneous emission from optical amplification. Unsupervised machine learning clustering (e.g., K-means) has recently been proposed as a practical approach to the blind compensation of stochastic and deterministic nonlinear distortions. In this work, the Density-Based Spatial Clustering of Applications with Noise (DBSCAN) algorithm is employed, for the first time, for blind nonlinearity compensation. DBSCAN is tested experimentally in a 40 Gb/s 16 quadrature amplitude-modulated system at 50 km of standard single-mode fiber transmission. It is shown that at high launched optical powers, DBSCAN can offer up to 0.83 and 8.84 dB enhancement in Q-factor when compared to conventional K-means clustering and linear equalisation, respectively.

Keywords: fiber optics communications; coherent communications; machine learning; clustering; nonlinearity cancellation

1. Introduction

Coherent optical communications have been proposed as a viable solution for maximising the signal capacity in both short-reach and long-haul communications [1]. However, Kerr-induced fiber nonlinearity prevents channel capacity from approaching the Shannon limit, especially when the signal power is high [2]. Endeavours to surpass the Kerr nonlinearity limit have been performed by techniques that in principle compensate deterministic nonlinearities [3]. For example, nonlinearities can be combated by either inserting an optical phase conjugator (OPC) at the middle point of the link [4], or by inverting the fiber effects among multiple frequency stabilised optical signals [5]. However, OPC reduces the flexibility in an optically routed network, whereas in [5], a digital back propagation (DBP) [6] pre-compensator was used, which is of excessive complexity. Other famous techniques include hybrid pre- and post-compensation [7], Volterra-based nonlinear equalisation (NLE) [8], phase-conjugated twin waves (PC-TW) [9], and the nonlinear Fourier transform (NFT) [10]. Unfortunately, pre-/post-compensation algorithms and Volterra-NLE present marginal performance enhancement, PC-TW sacrifices signal capacity and NFT is unpractical for real-time signal processing.

Above all, the aforementioned deterministic methods are unable to tackle stochastic nonlinearities, such as the amplified spontaneous emission (ASE) noise induced from optical amplifiers.

Unsupervised machine learning clustering has been recently introduced in optical communications for blind (training data-free) nonlinear equalisation (BNLE). Such unsupervised algorithms can tackle stochastic nonlinearities and include, for example, fuzzy logic C-means [11], K-means [11,12], hierarchical [11], affinity propagation [13], and Gaussian mixture [14] clustering. The algorithms used in [11–14] are mainly used for compensating fiber nonlinearity in long-haul transmission systems, in which the distorted received constellation diagrams contain Gaussian-circular clusters. However, there is an uncertainty if the machine learning clustering algorithms can be effective for non-circular rotated clusters caused by the strong nonlinear phase noise.

In this work, the aforementioned issue is addressed by experimentally demonstrating the first BNLE that harnesses the Density-Based Spatial Clustering of Applications with Noise (DBSCAN) [15] algorithm in 40 Gb/s 16 quadrature amplitude modulation (QAM) coherent optical signals being transmitted at 50 km. As a proof of concept, DBSCAN is tested for very high launched optical powers (LOPs), where the clusters of the received constellation diagrams are vastly rotated by means of self-phase modulation (SPM). Two modified DBSCAN methods are also adopted, in which the "un-clustered" noisy points are further processed using (1) K-means and (2) the minimum distance between an unlabelled point and the clustered points. It is shown that DBSCAN offers up to 0.83 dB Q-factor improvement over K-means and 8.84 dB when compared to linear equalisation at +16 dBm of LOP. This occurs because DBSCAN can effectively recover non-circularly-symmetric (elliptical form) noisy clusters by effectively combating SPM.

2. Density-Based Spatial Clustering of Applications with Noise (DBSCAN) Description

In density-based clustering, an assumption is made: clusters are dense regions in space, separated by regions of lower density [16]. A dense cluster is a region which is "density connected", i.e., the density of points in that region is greater than a minimum [17]. DBSCAN is an example that searches for dense areas and expands these recursively to find arbitrarily dense-shaped clusters. The two main parameters of DBSCAN are the ε ("Epsilon") and the "minimum points". The ε defines the radius of the "neighbourhood region" while the "minimum points" define the minimum number of constellation points (i.e., symbols) that should be contained within that neighbourhood. DBSCAN arbitrarily picks up a point until all of them have been visited. If the predefined number of "minimum points" is within the radius ε, then all these points are considered to be part of the same cluster. The clusters are then expanded by recursively repeating the neighbourhood calculation for each neighbouring point. However, for the unallocated points, if the number of points within the ε-neighbourhood is less than a predefined threshold, they are designated to be "noisy" and are not assigned to a particular cluster. Noisy data are not further processed in conventional DBSCAN. Here, a 2nd loop clustering is proposed to be applied only for these noisy data using (1) K-means [11] or (2) the minimum distance between an unlabelled point and the clustered points. A schematic diagram for conventional DBSCAN is depicted in Figure 1 when the minimum points are 4. In Figure 1, the following assumptions are assumed [18]:

a. Epsilon neighbourhood (Nε): A set of all constellation points within a distance ε.
b. Core point: A constellation point whose Nε contains at least a "minimum point" (including itself).
c. Direct Density Reachable: A point q is directly density reachable from a point p, if p is a core point and $q \in$ Nε.
d. Density Reachable: Two constellation points (p, t) are density reachable if there is a chain of "direct density reachable" points that link these two points (p, q, t).
e. Border Point: A constellation point that is "direct density reachable" but not a core point.
f. Noise: Constellation points not belonging to any point's Nε.

The steps related to the conventional and modified DBSCAN are listed below, where the algorithm converges until all constellation points have been allocated to a cluster or labelled as "noisy" only if conventional DBSCAN is considered (step 5 below—1st loop) [17,18]:

1. Randomly select a point p (referred in Figure 1) in the constellation map.
2. Retrieve all constellation points directly density reachable from p that satisfy the condition of the radius ε limits.
3. If the constellation point p is a core point, a cluster is formed. Search recursively and find all of its density-connected points and assign them to the same cluster as p.
4. If p is not a core point, the DBSCAN algorithm "scans" for the rest of the unvisited constellation points.
5. DBSCAN 1st loop: Points that are un-clustered are labelled as zero points ("noisy points") where linear equalisation is performed only on these points, and then the conventional DBSCAN algorithm stops.
6. DBSCAN 2nd loop (extra novel step):

i. *Method-1:* K-means clustering is activated for the "noisy points" using the Lloyd's algorithm [11,12]:

a. Assignment: Allocate each observation to the cluster whose mean has the least squared Euclidean distance ("nearest" mean) [11,12].
b. Update: Calculate the new means to be the centroids of the observations in the new clusters [8,9]. K-means converges when assignments do not change.

ii. *Method-2:* Calculation of minimum distance between the unlabelled "noisy points" and the clustered points.

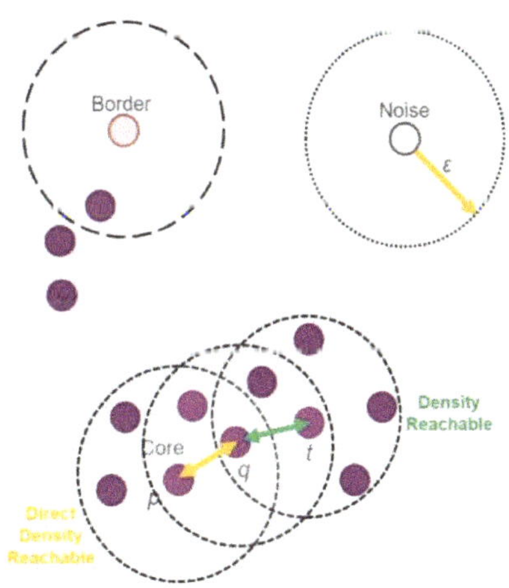

Figure 1. Density-Based Spatial Clustering of Applications with Noise (DBSCAN) example for Min. Points = 4.

3. Experimental Setup

Figure 2 depicts the schematic diagram of the experimental setup of the 10 GBaud (40 Gb/s) 16 QAM coherent signal. In the transmitter-digital signal processing (DSP), look-up table-based pre-distortion was used to mitigate the opto-electronic component impairments, similarly to [19]. A narrow linewidth (<100 kHz) external cavity laser (ECL) was tuned to 1549.5 nm and, using an arbitrary waveform generator (AWG) operating at 20 GS/s, two uncorrelated pseudo-random level signals ($2^{15}-1$) were applied to the in-phase/quadrature (I/Q) modulator to generate the 16 QAM signal. After IQ modulation, the optical signal was transmitted over 50 km of standard single-mode fiber (SSMF). At the receiver, noise loading was added using an optical amplifier to set different optical signal-to-noise ratio (OSNR) values and subsequently the optical signal was converted to an electrical one using a homodyne coherent receiver. Afterwards, the signal was captured by a real-time oscilloscope sampled at 50 GS/s for offline receiver-DSP, in which the data was first resampled to two samples/points using prior knowledge of the clock frequency. Then the constant modulus algorithm (CMA) combined with the multi-modulus algorithm (MMA) was utilised for signal equalisation. An Mth power frequency drifting compensation method was employed to compensate the frequency offset between the signal and the local oscillator in the coherent receiver. The decision-directed phase-locked loop (DDPLL) method was employed for the carrier phase recovery. Finally, the machine learning algorithm was processed before the hard decision and the bit error rate (BER)/Q-factor (=$20\log_{10}\left[\sqrt{2}erfc^{-1}(2BER)\right]$) calculation, similarly to other reported work with machine learning signal processing [20–25].

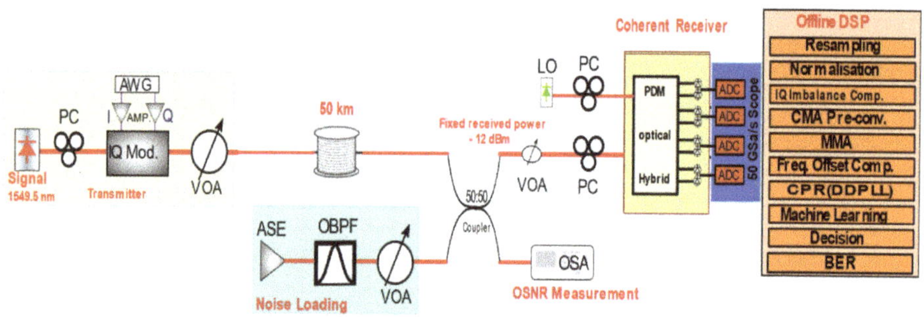

Figure 2. Experimental setup for a 40 Gb/s 16 quadrature amplitude modulation (QAM) coherent optical signal transmitted at 50 km, incorporating machine learning clustering. PC: polarisation controller, OBPF: optical band-pass filter, LO: local oscillator, CMA/MMA: constant/multi-modulus algorithm, CPR (DDPLL): carrier phase recovery (decision-directed phase-locked loop).

4. Results

A 40 Gb/s 16 QAM waveform is transmitted with +16 dBm LOP over 50 km. Two parameters are needed to optimise the DBSCAN algorithm to produce the lowest BER, namely ε and the minimum number of points. It is worth noting that in a real-time communication link, the optimisation process of DBSCAN should be run offline as a training stage. In this stage, the optimal Voronoi regions would be generated, where in the real-time processing the incoming data would be assigned directly to ideal symbols. It is envisaged that such an approach should be very effective in optical communication links, where linear and nonlinear effects remain relatively stable over time, and therefore the training process should be run only once. The calculated BER while scanning for ε and the minimum number of points is shown in Figure 3. From Figure 3, it is evident that the majority of the best BER values can be found for $0.1 < \varepsilon < 0.45$. For these values of ε, the minimum points do not affect the BER much except when $0.1 < \varepsilon < 0.14$, where the minimum points should be less than 110. In Figure 4, the performance of clustering algorithms is shown for different LOPs and two values of received OSNR: 30 and 15 dB. In Figure 4, the performance benefit of machine learning clustering over linear equalisation is significant for both OSNR values, especially when using DBSCAN method-2, resulting in up to 8.8 dB Q-factor improvement. This is attributed to the compensation of SPM since single-channel transmission is carried out. Results indicate that DBSCAN-based BNLE is a robust soft-clustering method when very strong nonlinear phase noise is present and where linear equalisation fails completely. Moreover, DBSCAN method-2 has the highest Q-factor along the whole range of LOPs. Compared to DBSCAN method-1 and K-means, method-2 increases the Q-factor by up to about 0.7 and 0.83 dB, respectively, by better handling highly rotated clusters that become almost elliptically shaped. This is because the overlapping (soft) clustering ability of method-2 is more powerful than the common hard (exclusive) clustering of K-means and method-2 (which also includes K-means for the noisy constellation points). This is confirmed by the received constellation diagrams of Figure 5b, related to +16 dBm of LOP (OSNR = 30 dB). On the other hand, the quite similar performances between DBSCAN algorithms and K-means at lower LOPs and OSNR is due to the existence of nearly circular Gaussian clusters, which are not rotated. This is corroborated in the received 16 QAM constellation diagrams of Figure 5a at +10 dBm of LOP (OSNR = 15 dB). In the left constellation diagram of Figure 5a, the DBSCAN "noisy" points are also presented in the 1st loop of the algorithm.

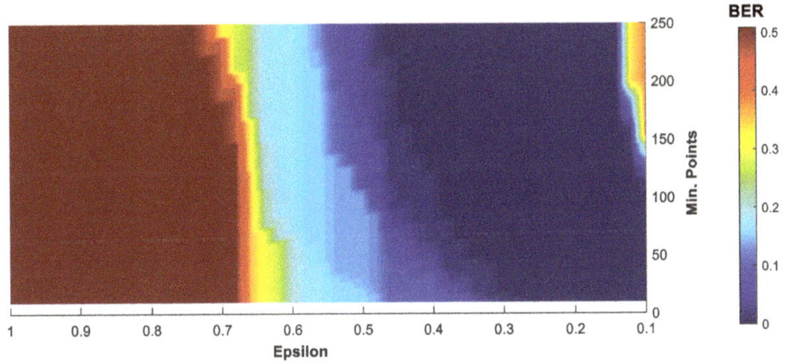

Figure 3. DBSCAN optimisation for 16 QAM transmission over 50 km at +10 dBm of launch power: bit-error-rate (BER) vs. ε ("Epsilon"), Min. Points = 4.

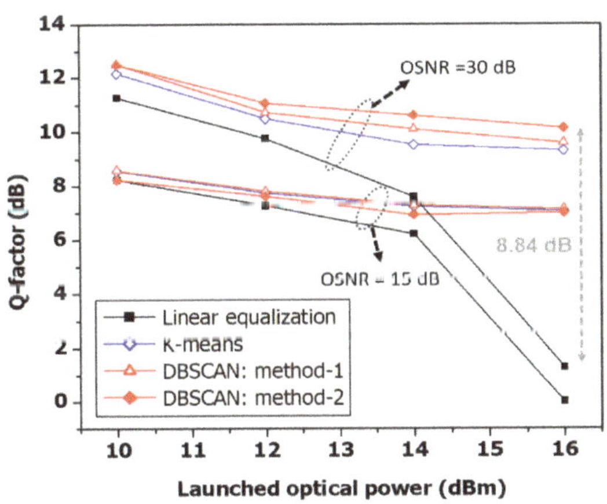

Figure 4. DBSCAN vs. K-means for 16 QAM transmission at 50 km for different launched optical powers (LOPs) when optical signal-to-noise ratio (OSNR) is 30, 15 dB.

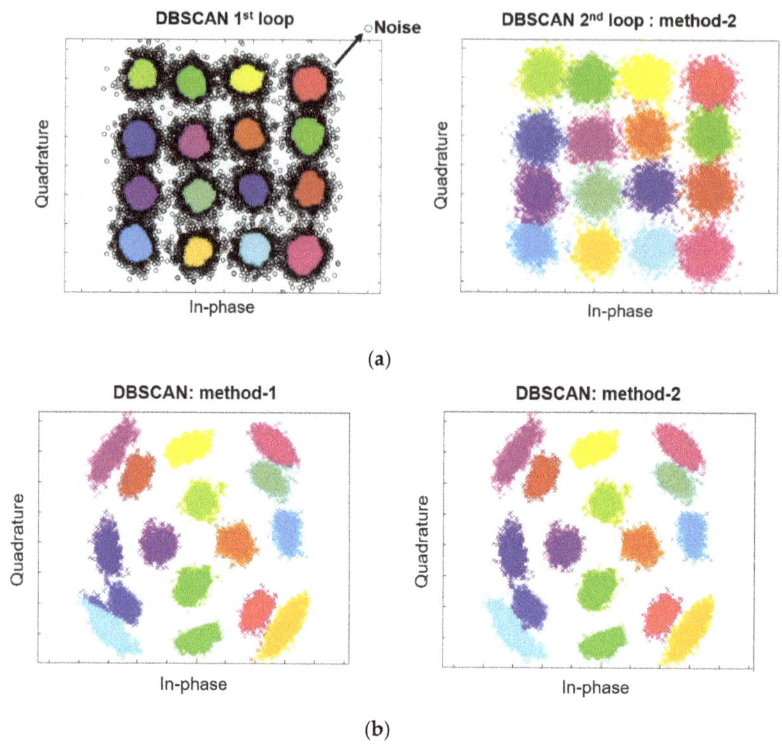

Figure 5. Received constellation diagrams for (**a**) DBSCAN 1st loop (left), method-2/2nd loop (right) at +10 dBm of LOP (OSNR = 15 dB); and (**b**) DBSCAN method-1 (left), method-2 (right) at +16 dBm of LOP (OSNR = 30 dB).

5. Conclusions

The first DBSCAN-BNLE was experimentally demonstrated for 16 QAM at 50 km. Two novel DBSCAN methods were proposed, in which the "un-clustered" noisy constellation points were processed using (1) K-means and (2) the minimum distance between an unlabelled point and the clustered points. Compared to linear equalisation, method-2 improved the Q-factor up to 8.8 dB by combating SPM. Method-2's ability for overlapping clustering resulted in Q-factor improvement over method-1 and K-means (exclusive clustering), when vastly rotated clusters of nearly elliptical form occur. Once optimised, DBSCAN proved to be a robust BNLE for very strong nonlinear phase noise.

In future work, inter-channel nonlinear effects, such as four-wave mixing and cross-phase modulation, will be tackled using the proposed machine learning algorithm in a super-channel transmission scenario that incorporates many wavelength channels.

Author Contributions: Conceptualization, E.G. and Y.L.; methodology, E.G.; software, E.G. and Y.L.; validation, M.J., S.O., K.M., P.F.W. and L.P.B; formal analysis, M.J.; investigation, E.G.; resources, L.P.B.; data curation, K.M. and P.F.W.; writing—original draft preparation, E.G. and Y.L.; writing—review and editing, E.G., Y.L., S.O., and L.P.B; visualization, E.G., Y.L. and S.O.; supervision, L.P.B.; project administration, E.G. and L.P.B.; funding acquisition, E.G. and L.P.B.

Funding: This work was supported by the Science Foundation Ireland through grant numbers 13/RC/2077, 12/RC/2276, 15/US-C2C/I3132, the HEA INSPIRE Programme, and the EU/EDGE Marie Curie programme with grant number 713567.

Conflicts of Interest: The authors declare no conflict of interest. The funders had no role in the design of the study; in the collection, analyses, or interpretation of data; in the writing of the manuscript, nor in the decision to publish the results.

References

1. Chagnon, M. Optical Communications for Short Reach. *J. Lightwave Technol.* **2019**, *37*, 1779–1797. [CrossRef]
2. Ryu, S.; Yamamoto, S.; Taga, H.; Edagawa, N.; Yoshida, Y.; Wakabayashi, H. Long-haul coherent optical fiber communication systems using optical amplifiers. *J. Lightwave Technol.* **1991**, *9*, 251–260. [CrossRef]
3. Ellis, A.D.; Zhao, J.; Cotter, D. Approaching the nonlinear Shannon-limit. *J. Lightwave Technol.* **2010**, *28*, 423–433. [CrossRef]
4. Al-Khateeb, M.A.Z.; McCarthy, M.E.; Sánchez, C.; Ellis, A.D. Nonlinearity compensation using optical phase conjugation deployed in discretely amplified transmission systems. *Opt. Express* **2018**, *26*, 23945–23959. [CrossRef]
5. Temprana, E.; Myslivets, E.; Kuo, B.P.; Liu, L.; Ataie, V.; Alic, N.; Radic, S. Overcoming Kerr-induced capacity limit in optical fiber transmission. *Science* **2015**, *348*, 1445–1448. [CrossRef]
6. Maher, R.; Galdino, L.; Sato, M.; Xu, T.; Shi, K.; Kilmurray, S.; Savory, S.J.; Thomsen, B.C.; Killey, R.I.; Bayvel, P. Linear and nonlinear impairment mitigation in a Nyquist spaced DP-16QAM WDM transmission system with full-field DBP. In Proceedings of the ECOC, Cannes, France, 21–25 September 2014; IEEE: Piscataway, NJ, USA, 2014; p. 5.10.
7. Lowery, A.J. Fiber nonlinearity pre-and post-compensation for long-haul optical links using OFDM. *Opt. Express* **2007**, *15*, 12965–12970. [CrossRef]
8. Giacoumidis, E.; Aldaya, I.; Jarajreh, M.A.; Tsokanos, A.; Le, S.T.; Farjady, F.; Jaouën, Y.; Ellis, A.D.; Doran, N.J. Volterra-based reconfigurable nonlinear equalizer for dual-polarization multiband coherent OFDM. *IEEE Photonics Technol. Lett.* **2014**, *26*, 1383–1386. [CrossRef]
9. Liu, X. Phase-conjugated twin waves for communication beyond the Kerr nonlinearity limit. *Nat. Photonics* **2013**, *7*, 560–568. [CrossRef]
10. Le, S.T.; Aref, V.; Buelow, H. Nonlinear signal multiplexing for communication beyond the Kerr nonlinearity limit. *Nat. Photonics* **2017**, *11*, 570–576. [CrossRef]
11. Giacoumidis, E. Blind Nonlinearity Equalization by Machine Learning based Clustering for Single-and Multi-Channel Coherent Optical OFDM. *IEEE J. Lightwave Technol.* **2018**, *36*, 721–727. [CrossRef]
12. Zhang, J.; Chen, W.; Gao, M.; Shen, G. K-means-clustering-based fiber nonlinearity equalization techniques for 64-QAM coherent optical communication system. *Opt. Express* **2017**, *25*, 27570–27580. [CrossRef] [PubMed]
13. Giacoumidis, E. Affinity propagation clustering for blind nonlinearity compensation in coherent optical OFDM. In Proceedings of the CLEO, San Jose, CA, USA, 13–18 May 2018; OSA: Rochester, NY, USA, 2018; p. STh1C.5.
14. Zibar, D.; Piels, M.; Jones, R.; Schäeffer, C.G. Machine Learning Techniques in Optical Communication. *IEEE J. Lightwave Technol.* **2016**, *34*, 1442–1452. [CrossRef]
15. Ester, M.; Kriegel, H.P.; Sander, J.; Xu, X. A Density-Based Algorithm for Discovering Clusters in Large Spatial Databases with Noise. In Proceedings of the International Conference on Knowledge Discovery and Data Mining, Portland, OR, USA, 2–4 August 1996; pp. 226–231.
16. Shen, J. Real-Time Superpixel Segmentation by DBSCAN Clustering Algorithm. *IEEE Trans. Image Process.* **2016**, *25*, 5933–5942. [CrossRef] [PubMed]
17. Boada, R.; Borkowski, R.; Monroy, I.T. Clustering algorithms for Stokes space modulation format recognition. *Opt. Express* **2015**, *23*, 15521–15531. [CrossRef]
18. Lu, X. An I-Q-Time 3-dimensional post-equalization algorithm based on DBSCAN of machine learning in CAP VLC system. *Opt. Commun.* **2019**, *430*, 299–303. [CrossRef]
19. Zhang, J.; Yu, J.; Chien, H.C. Advanced linear and nonlinear compensations for 16QAM SC-400G unrepeatered transmission system. *Opt. Commun.* **2018**, *409*, 34–38. [CrossRef]
20. Nguyen, T.; Mhatli, S.; Giacoumidis, E.; Van Compernolle, L.; Wuilpart, M.; Mégret, P. Fiber nonlinearity equalizer based on support vector classification for coherent optical OFDM. *Photonics J.* **2016**, *8*, 1–9. [CrossRef]
21. Giacoumidis, E.; Mhatli, S.; Nguyen, T.; Le, S.T.; Aldaya, I.; McCarthy, M.E.; Ellis, A.D.; Eggleton, B.J. Comparison of DSP-based nonlinear equalizers for intra-channel nonlinearity compensation in coherent optical OFDM. *Opt. Lett.* **2016**, *41*, 2509–2512. [CrossRef]

22. Giacoumidis, E.; Mhatli, S.; Stephens, M.F.; Tsokanos, A.; Wei, J.; McCarthy, M.E.; Doran, N.J.; Ellis, A.D. Reduction of Nonlinear Inter-Subcarrier Intermixing in Coherent Optical OFDM by a Fast Newton-based Support Vector Machine Nonlinear Equalizer. *J. Lightwave Technol.* **2017**, *35*, 2391–2397. [CrossRef]
23. Giacoumidis, E.; Mhatli, S.; Le, S.T.; Aldaya, I.; McCarthy, M.E.; Ellis, A.D.; Eggleton, B.J. Nonlinear Blind Equalization for 16-QAM Coherent Optical OFDM using Support Vector Machines. In Proceedings of the ECOC, Düsseldorf, Germany, 18–22 September 2016; IEEE: Piscataway, NJ, USA, 2016; p. Th.2.P2.
24. Mhatli, S.; Mhatli, S.; Mrabet, H.; Dayoub, I.; Giacoumidis, E. A novel SVM robust model Based Electrical Equalizer for CO-OFDM Systems. *IET Commun.* **2017**, *11*, 1091–1096. [CrossRef]
25. Giacoumidis, E.; Tsokanos, A.; Ghanbarisabagh, M.; Mhatli, S.; Barry, L.P. Unsupervised Support Vector Machines for Nonlinear Blind Equalization in CO-OFDM. *Photonics Technol. Lett.* **2018**, *30*, 1091–1094. [CrossRef]

applied sciences

MDPI

Article

Reduced-Complexity Artificial Neural Network Equalization for Ultra-High-Spectral-Efficient Optical Fast-OFDM Signals

Mutsam A. Jarajreh

Computer Engineering Department, Fahad Bin Sultan University, Tabuk 71454, Saudi Arabia; mjarajra@fbsu.edu.sa

Received: 30 July 2019; Accepted: 16 September 2019; Published: 27 September 2019

Abstract: Digital-based artificial neural network (ANN) machine learning is harnessed to reduce fiber nonlinearities, for the first time in ultra-spectrally-efficient optical fast orthogonal frequency division multiplexed (Fast-OFDM) signals. The proposed ANN design is of low computational load and is compared to the benchmark inverse Volterra-series transfer function (IVSTF)-based nonlinearity compensator. The two aforementioned schemes are compared for long-haul single-mode-fiber-based links at 9.69 Gb/s direct-detected optical Fast-OFDM signals. It is shown that an 80 km extension in transmission-reach is feasible when using ANN compared to IVSTF. This occurs because ANN can tackle stochastic nonlinear impairments, such as parametric noise amplification. Using ANN, the dynamic parameters requirements of the sub-ranging quantizers can also be relaxed compared to linear equalization, such as the reduction of the optimum clipping ratio and quantization bits by 2 dB and 2-bits, respectively, and by 2 dB and 2 bits when compared to the IVTSF equalizer.

Keywords: optical Fast-OFDM; neural networks; nonlinearity compensation; optical fiber communications

---

1. Introduction

As one of the most dominant high spectral-efficiency methods, optical-orthogonal frequency division multiplexing (O-OFDM) can virtually eliminate the interference among received symbols induced by fiber dispersion and the effect of random polarization rotation [1]. In OFDM equally MHz spaced parallel subcarriers are used to form low capacity data transmission signals [2]. However, Rodrigues and Darwazeh recently proposed an alternative ultra-high spectral-efficient scheme, the Fast-OFDM [3], which harnesses the compression properties of Inverse Fast Cosine Transform (IFCT). In Fast-OFDM the frequency spacing between sub-carriers is considerably decreased, resulting in increased bandwidth efficiency compared to the traditional O-OFDM. This occurs because Fast-OFDM uses only half of the sub-carrier spacing [3]. Though, sub-MHz sub-carrier spacing produces significant distortion in phase-modulated signals, thus it is essential to apply single-dimensional modulation formats such as amplitude shifted-keying (ASK) [3]. IFCT-based optical Fast-OFDM has been previously demonstrated for long-haul coherent optical double-side band signals [4,5]. Furthermore, direct-detected optical Fast-OFDM was employed for cost-sensitive local networks that make use of multimode fiber links [6], due to the high cost-efficiency of the technique [5]. However, similarly to conventional O-OFDM, the Fast-OFDM signals also suffer from a high peak-to-average power ratio resulting in similar transmission performance with O-OFDM under the same signal capacity [6]. In the same work from Giacoumidis et al. [6], the dynamic parameters requirements of the sub-ranging quantizers (well-identified as digital-to-analogue and analogue-to-digital converters, DACs/ADCs [7]) involved in optical Fast-OFDM signals were also analyzed as a proof-of-concept for future-proof real-time implementation. On the other hand, while linear equalization has been an easy task for both OFDM and Fast-OFDM signals, fiber-induced nonlinear distortion compensation has not been properly addressed in optical Fast-OFDM. In traditional O-OFDM, several digital signal processing (DSP) techniques for mitigation of fiber nonlinearities have been proposed, such as the digital back propagation (DBP) [8], the inverse Volterra-series transfer function (IVSTF) [9], and hybrid pre- and post-nonlinearity compensators [10,11]. In DBP the equalizer attempts to reverse the channel non-linear effects, where the SMF channel is thoroughly and extensively modelled, thereafter, the received signals are digitally back-propagated over the modelled channel with the help of the split-step Fourier (SSF) operations at a very small distances may be 40 FFT/IFFT operations per span; which makes this method impractical and computationally expensive for real-time applications as there are a huge number of computation steps [12,13]. On the other hand, The IVSTF algorithm was presented in order to reduce the complexity of digital back-propagation (DBP); this removed the necessity for the split-step Fourier (SSF) method, which is computationally incompetent. The VSTF offers an analytical tool for expressing the fiber non-linear effects by similarly constructing the inverse channel based on IVSTF depending upon the number of fiber spans not the fiber length as in DBP; which significantly reduces computational complexity compared to DBP [12–15]. Moreover, IVSTF has shown marginal signal quality-factor improvement for coherent O-OFDM using 16-quadrature amplitude modulated (16-QAM) sub-carriers but at lower DSP computational effort [11]. Very recently, artificial neural network (ANN), which is a mimic of the conceptual and behavioral nature of the biological networks neural networks, have proven abilities and have been applied to various fields such as, competitive learning [16], finance [17], control system [18], energy management [19], and telecommunications where it has been recently used as an equalizer for coherent O-OFDM signals revealing promising results for long-haul links [14].

Due to the very low frequency spacing between sub-carriers, Fast-OFDM signals suffer more from inter-carrier interference compared to the conventional O-OFDM [3,15], consequently, the importance of realizing an equalizer to mitigate nonlinear impairments in optical Fast-OFDM is much higher. An attempt to tackle nonlinearities in optical Fast-OFDM systems was made via a Wiener-Hammerstein electrical equalizer, with however marginal performance improvement [20]. Unfortunately, the use of ANN nor IVSTF has not been reported yet, therefore in this work, for the first time, a low-complex ANN-based machine learning NLE is numerically demonstrated in low-cost intensity-modulated and directed-detected optical Fast-OFDM links using a standard single-mode fiber (SMF) as a transmission medium. A comparison is also made with the deterministic IVSTF which is also implemented for the first time in optical Fast-OFDM. The paper is organized as follows: Section 2 shows the linear equalization performance of the direct-detected-based optical Fast-OFDM system over additive white Gaussian noise (AWGN). In Section 3, there are details on the adopted ANN algorithm and IVSTF for optical Fast-OFDM signals. In Section 4, ANN is tested over an SMF-link, while in Section 5, a comparison between ANN and IVSTF is conducted for Fast-OFDM signals over SMF, considering the impact of key DAC/ADC parameters when compared to linear equalization. Thereafter in Section 6, a detailed computational complexity analysis of the ANN-NLE and IVSTF-NLE is conducted in terms of number of subcarriers for different system parameter is discussed. Finally, the paper is concluded in Section 7.

2. Impact of Directed-Detected Optical Fast-OFDM Signals over AWGN Using Linear Equalization

For our numerical simulation, the direct-detected optical Fast-OFDM transceiver design is similar to [6], using 64 sub-carriers and ASK formats for modulation with 4, 8, and 16 amplitude levels [6]. The adopted sampling rate was taken at 6.25 GS/s, resulting in sub-carrier frequency bandwidth of about 98 MHz and a length per symbol of 12.8 ns, out of which 3.2 ns were wasted by the insertion of a cyclic prefix (CP) to avoid interference among symbols during transmission.

In Figure 1, the optical Fast-OFDM modem performance over an AGWN channel is illustrated in terms of optical signal-to-noise ratio (OSNR) at 0.1 nm without using any form of nonlinear equalization. In Figure 1, the DAC/ADC clipping noise (or ratio) and quantization bits were isolated to investigate the true impact of OSNR on the developed optical Fast-OFDM modem. From Figure 1, it is noticed that for a fixed bit-error-rate (BER) value, a higher order modulation format needs higher OSNR. More specifically, the OSNR is required for achieving a BER of 10^{-3} when using 4-ASK is 18.5 dB, 25.5 dB for 8-ASK, and 36 dB for 16-ASK. It should be noted that simulated results agree very well with the analytical results reported in [6], which confirms the validity of the used model.

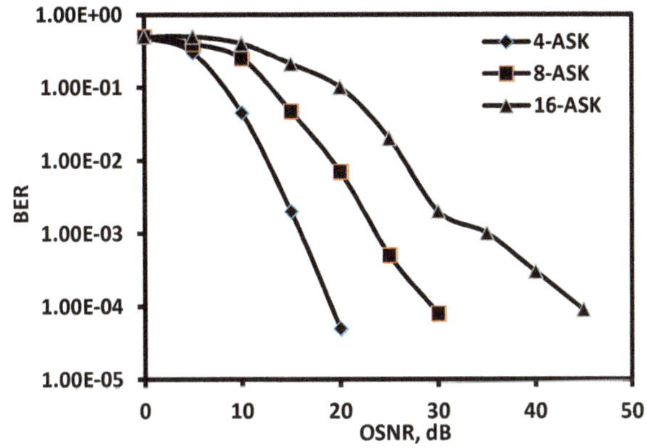

Figure 1. Optical signal-to-noise ratio (OSNR) vs. bit-error-rate (BER) performance of optical Fast-orthogonal frequency division multiplexing (Fast-OFDM) over an additive white Gaussian noise (AWGN) channel.

3. ANN and IVSTF Nonlinear Equalizers

ANN using nonlinear decision boundaries [21] via the multilayer perceptron has been considered as a promising alternative for combating impairments in wireless communications [22,23]. In ANN, complex mapping between input and output spaces could be achieved, however, at the cost of complexity [14,24]. The proposed ANN for ASK-based optical Fast-OFDM signals reduces the complexity of the equalizer since it processes only amplitude levels and not phase data nor complex data. A schematic diagram is depicted in Figure 2b, in which $s(k)$ FOFDM denotes the training transmitted Fast-OFDM sub-carriers, k (FOFDM refers to Fast-OFDM). The developed NLE using ANN consists of k neural-sub-networks in which the inherent neurons are interconnected with the Fast-OFDM weights values $wFOFDM$ (k,i). The number of neurons in every neural sub-net is proportional to the level of modulation format; for example, it is 16 for 16-ASK. In the output layer, the output signal sums-up all Fast-OFDM neurons $\hat{s}(k)_{FOFDM}$. To calculate the error, the minimum mean-square error (MMSE) is estimated, after which the neurons/weights are updated similarly to an adaptive digital filter [14]. Such learning process continues until a very low error value is reached (convergence rate). Equation (1) shows how this error is estimated:

$$e(k)_{FOFDM} = s(k)_{FOFDM} - \hat{s}(k)_{FOFDM}, \tag{1}$$

In Equation (1), $\hat{s}(k)_{FOFDM}$ is the signal after the training stand can be calculated by the following mathematical formula:

$$\hat{s}(k)_{FOFDM} = \sum_{i=1}^{16} w_{FOFDM}(k,i)\varphi_{FOFDM}(k,i)(s(k)_{FOFDM}), \tag{2}$$

In Equation (2), $\varphi_{FOFDM}(k,i)$ represents a nonlinear sigmoid "activation function" [14]. It should be also noted that the total number of neurons adopted is also proportional to the number of sub-carriers, as the Fast-OFDM signal consists of (k) number of subcarriers; thus, the ANN equalizer consists of (k) sub-neural networks, each sub-network is designated for a different subcarrier.

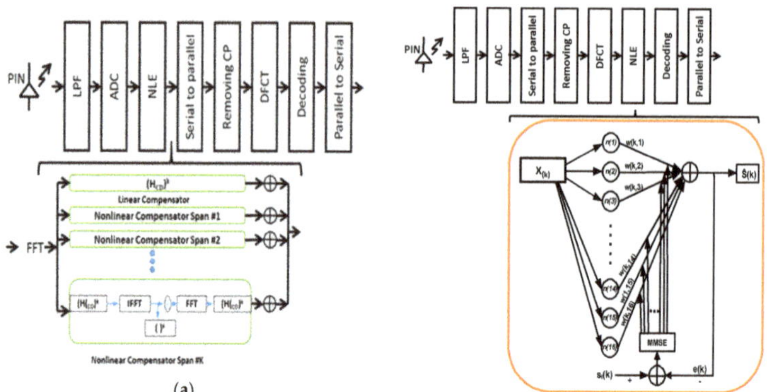

(a) (b)

Figure 2. (**a**) Optical Fast-OFDM received diagram showing the equalizer that is based on IVSTF [11]. (**b**) 16-ASK optical Fast-OFDM received diagram illustrating the ANN sub-neural network equalizer. LPF: low-pass filter; ADC: analogue-to-digital converter; CP: cyclic prefix; DCT: Discrete-Cosine Transform; NLE: nonlinear equalizer; ANN: artificial neural network; n: neuron; MMSE: minimum mean square error; HCD: nonlinear system chromatic dispersion.

The designed ANN for NLE operation in optical Fast-OFDM performs Riedmiller-resilient based back-propagation (RP) [24–27] for the purpose of updating the weights. The steepest descent algorithm is also used to find a global-data minimization [14]. In ANN numerical demonstration, a differentiable hyperbolic tangent function was employed for the hidden layer (which is only one to further reduce complexity) [26,27] and the linear function "purelin" for the output layer. In Figure 2b the MMSE is processed via the RP algorithm in order for ANN to find the weights that minimize the error in terms of the vectors $S(n)$ *FOFDM* and $\hat{S}(n)$ *FOFDM*:

$$E(n)_{FOFDM} = \| S(n)_{FOFDM} - \hat{S}(n)_{FOFDM} \|^2, \tag{3}$$

In the developed NLE using ANN the weights are updated according to the five steps presented in [14,26]. Figure 2a, illustrates the benchmark received IVSTF based NLE with identical procedure to [11], which is of deterministic nature since it can tackle deterministic linear and nonlinear crosstalk effects. IVSTF was used as a comparative NLE and it was implemented in time domain in contrast to ANN-NLE, which is processed in frequency domain after the DFCT block. It should be noted that the Volterra-based equalizer is of reduced complexity compared to the well-known full-step digital back propagation [11], using up to 3rd order kernels to account for up to 2nd order chromatic dispersion. The complexity, however, of IVSTF is much higher than ANN since it consists of multiple FFT & IFFT blocks for fiber dispersion and nonlinearity compensation, respectively.

4. Direct-Detected Optical Fast-OFDM System Model Equipped with NLEs and Performance over SMF

The proposed direct-detected optical Fast-OFDM system equipped with either ANN or IVSTF was implemented in a hybrid MATLAB with VPI® simulation platform. Similar equalizers have been previously employed in [11,14], which validates the model used in this work. Moreover, the Fast-OFDM system was modulated using 4-, 8- and 16-ASK signal formats at a signal capacity of 9.69, 14.53 and 19.3 Gb/s, respectively, for both equalizers using 64 sub-carriers and 1000 Fast-OFDM symbols. The transmission-reach of the developed system was set at 640 km (8 spans with a fiber-link length of 80 km). The length of CP was set at 25% to ensure effective elimination of inter-symbol interference [28,29]. At the receiver-end of the simulation set-up a single low pass filter (LPF) unit was employed having 3 dB bandwidth of approximately 3 GHz. For optical amplification per span, Erbium-doped fiber amplifiers (EDFAs) were used having a realistic noise figure of 6 dB.

The sampling rate of the DAC at the transmitter and the ADC at the receiver was chosen at 6.25 GS/s, while the ratio of clipping and quantization bits were optimally set to 7 dB and 10, respectively, which has been reported as the optimum values for high level modulation formats in O-OFDM [6]. The optimum launched optical power for our system under test was found at −6 dBm. A single unit positive-intrinsic-negative (PIN) photo-detector with 0.9 A/W of responsivity was adopted at the receiver-side for direct-detection, while the SMF-transmission-link had the following parameters: 16.9 ps/nm/km fiber dispersion, 0.21 dB/km fiber loss, 0.07 ps/km/nm^2 dispersion slope, 0.11 ps/km 0.5 polarization-mode-dispersion coefficient, 2.69 × 10-20 m^2/W Kerr-induced nonlinearity coefficient, and 80 μm^2 effective core area. The key simulation parameters for the developed ANN-based Fast-OFDM transmission model are depicted in Table 1.

Table 1. Key optical Fast-OFDM transceiver parameters.

Parameter	Description & Unit
Signal modulation format	4-, 8-, 16-ASK
Signal data bit-rate	9.69, 14.53, 19.37 Gb/s
Operating wavelength	1550 nm
Number of sub-carriers	64
Cyclic prefix (CP) length	25%
Forward-error-correction	7%
ANN training vector length	5%
Photo-detector type	PIN
PIN sensitivity	−19 dBm
DAC/ADC sampling rate	6.25 GS/s
DAC/ADC quant. bits	7
DAC/ADC clipping ratio	13 dB
LPF roll-off function	Bessel-Thomson
LPF 3 dB bandwidth (order)	3 GHz (2nd order)
EDFA gain (noise figure)	16 dB (6 dB)
SSMF span (length)	8 (80 km)

The transmission performance of the direct-detected optical Fast-OFDM system over SMF links for different modulation formats is shown in Figure 3. It is evident from this figure that for a fixed BER at 10^{-3} (the selected forward-error-correction limit at 10^{-9}) the maximum transmission-reach is at about 380 km, 240 km, and 200 km for 4-, 8- and 16-ASK modulation, respectively. This can be explained by the fact that a higher level of modulation that carry higher data rate increases the SNR on Fast-OFDM sub-carriers, due to the increased amplitude distortions induced by the combined effects of fiber chromatic dispersion and nonlinearity at higher transmission distances.

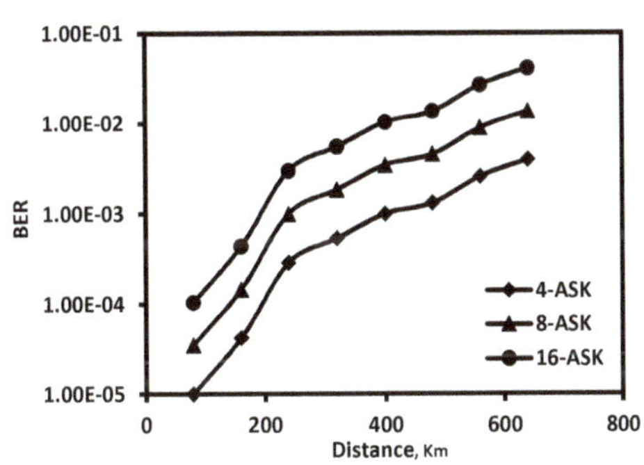

Figure 3. Optical Fast-OFDM transmission-reach in terms of BER.

5. ANN vs. IVSTF in Direct-Detected Optical Fast-OFDM

The capability of ANN for nonlinearity compensation in optical Fast-OFDM with intensity-modulation and direct-detection is illustrated in Figure 4 in terms of BER vs. distance. In Figure 4, a 4-ASK modulation was considered for simplicity at a signal capacity of 9.69-Gb/s. The optimum transmitted power was found at −6 dBm for ANN-NLE, as well for the benchmark IVSTF and without (W/O) using NLE (i.e., considering only linear equalization). It is revealed from Figure 4 that 160 km extension in transmission-reach is achieved by ANN compared to linear equalization for the adopted BER threshold. In the same manner, 80 km extension in transmission-reach is observed when using ANN compared to the Volterra-based NLE. The latter can be explained by the fact that while IVSTF compensates deterministic intra-channel nonlinearities, ANN can tackle both deterministic and stochastic nonlinearities, i.e., parametric noise amplification from concatenated EDFAs and the interplay between nonlinearity and polarization-mode-dispersion.

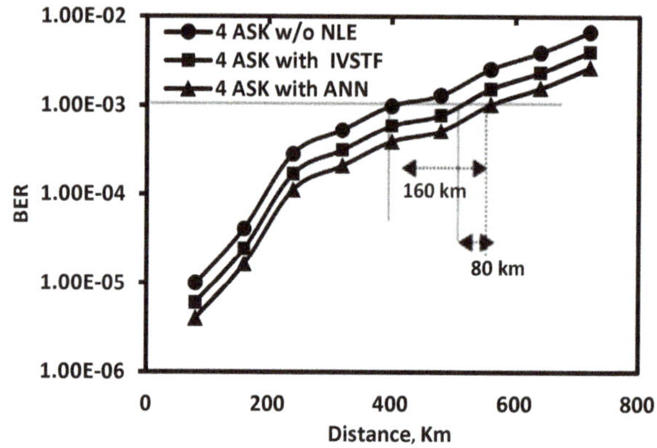

Figure 4. BER vs. transmission distance for 9.69-Gb/s optical Fast-OFDM at optimum launched optical power of −6 dBm for Volterra-series transfer function (IVSTF)/ANN-NLEs and without (W/O) using NLE.

On the other hand, the performance difference between ANN and linear equalization can be explained in Figure 5, where the BER distribution on the sub-carrier index is depicted. The figure essentially shows that a few Fast-OFDM sub-carriers under ANN equalization have decreased BER (corresponding to increased SNR) compared to linear equalization. This happens because ANN can reduce the effects of inter-carrier four-wave mixing.

Figure 6 depicts the received optical power at the PIN versus the transition distance while adopting the 4-ASK modulation format; it is worth noting that the optical received power values are different at different fiber lengths, varying from −7 dBm for 80 km to −18 dBm for 560 km.

Figure 7a shows the effect of the IVSTF and ANN equalizer on the optimum quantization bit. As it can be seen from this figure, utilizing the ANN equalizer relaxes the requirement of a high-resolution DAC/ADC, and reduces the requirement of high quantization bits by 2-bits for the same transmission distance (for the case of 580 km) when compared with case of without equalizer, and by 1 bit when compared with the case of utilizing the IVSTF equalizer. Similarly, this is also applied for the case of clipping ratio, as depicted in Figure 7b: using the established ANN equalizer, an increase in the transmission distance is observed when utilizing the optimum clipping ratio (7 dB); while for a fixed distance of 580 km, a 2-dB reduction in the required clipping ratio is observed when using ANN compared W/O ANN, moreover a reduction of 1 dB when compared with the IVSTF equalizer case.

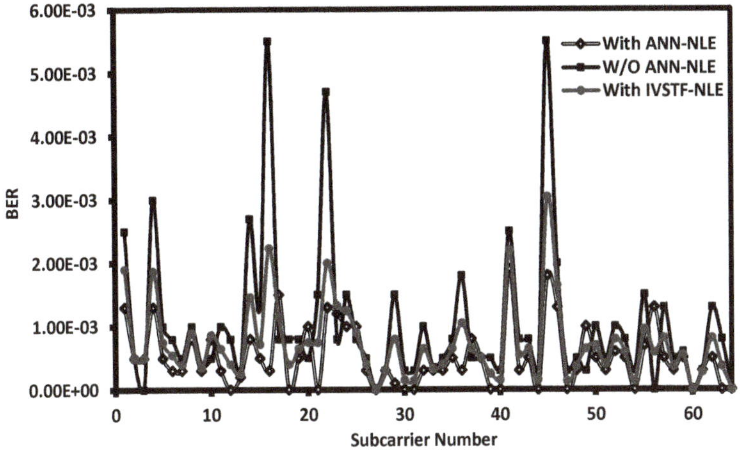

Figure 5. BER vs. optical Fast-OFDM sub-carrier index, with IVSTF-NLE, ANN_NLE and without utilizing ANN at 320 km of fiber transmission.

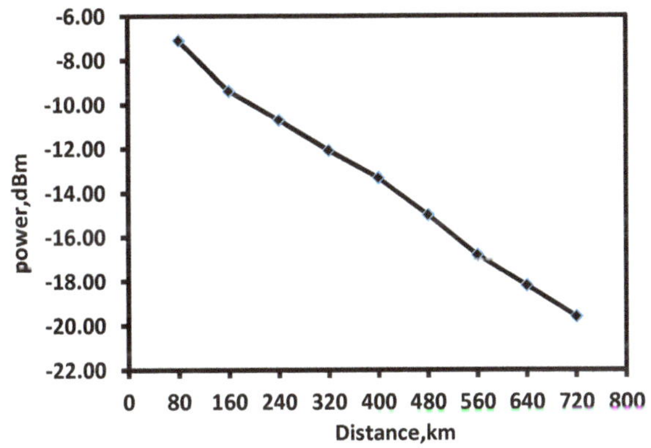

Figure 6. Received optical power vs. transmission distance for 9.69-Gb/s optical Fast OFDM without (W/O) using NLE.

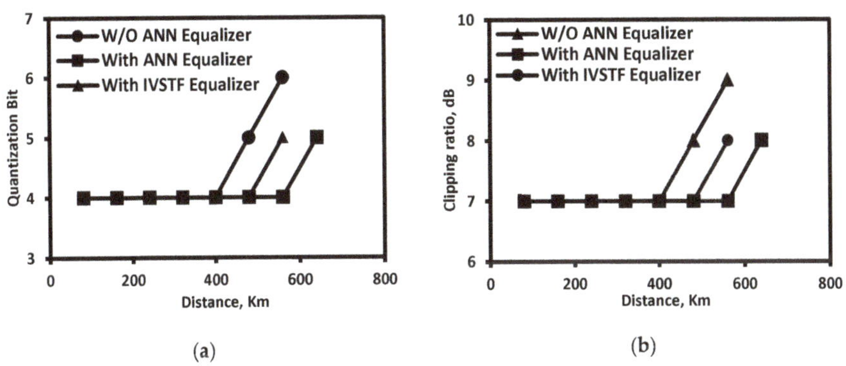

Figure 7. Effect of DAC/ADC components on 4-ASK optical Fast-OFDM transmission performance with the use of IVSTF equalizer, with ANN equalizer and W/O ANN at a signal bit rate of 9.69-Gb/s for a transmitted power of −6 dBm: (**a**) Quantization bit vs. Distance. (**b**) Clipping ratio vs. Distance.

6. Computational Complexity Analysis

When evaluating and comparing equalizers complexity it is essential to take the nature of the equalizer in to considerations, such as the DBP and IVSTF are essentially different from machine learning-based NLEs such as ANN and SVM. That is, as the IVSTF equalization mechanism is based on the idea of reversing of the effect propagation model; consequently, which makes it dependent on the number of fiber spans and the oversampling rate, therefore complexity, does not depend on other signal parameters such as modulation format levels [13].

As the IVSTF equalization is implemented in the hybrid time and frequency domain, it requires several conversions from time to frequency domain and vice versa, which intern makes the complexity depends on the FFT/IFFT pairs that operate on data blocks of size $Nblock = K * Nsub$, where k is the oversampling constant and $Nsub$ is the number of subcarriers. Consequently, the total number of operations for the IVSTF-based NLE is given by Equation (4), [13]:

$$N_{IVSTF} = \left(N_{Span} + 1\right) 8 N_{Block} Log_2 N_{Block} + \left(20 N_{Span} - 6\right) N_{Block} + 16\left(N_{Span} + 1\right), \quad (4)$$

On the contrary to IVSTF, ANN based NLEs complexity depends upon some signal parameters, such as the number of modulation forma level and the number of OFDM subcarriers.

The idea behind the ANN concept is based on the imitation of the human brain, as it utilizes a large number of low complexity neurons, which makes their implementation require fewer FLOPs when compared to other approaches like IVSTF. Therefore, the total number of operations performed for processing each Fast OFDM symbol can be obtained by Equation (5) [12,13].

$$FLOPS_{ANN} = Nop_{IL} + Nop_{HL} + Nop_{OL} = \left[N_{SC}{}^2 * M\right] + \left[(N_{SC} * M)^2 * (N_{HL} - 1)\right] + \left[N_{SC}{}^2 * M\right], \quad (5)$$

where the Nop_{IL}, Nop_{HL} and Nop_{OL} are the Flop count for input-, output-, and hidden-layers which are illustrated below in Equations (5)–(8) respectively.

$$Nop_{IL} = \left[N_{SC}{}^2 * M\right], \quad (6)$$

$$Nop_{HL} = (N_{SC} * M)^2 * (N_{HL} - 1), \quad (7)$$

$$Nop_{OL} = \left[N_{SC}{}^2 * M\right], \quad (8)$$

where N_{SC} is the number of subcarriers, M is modulation format level and N_{HL} is the number of hidden layers.

It is worth noting that in the Fast-OFDM case the complexity is reduced by a factor of two when compared to the conventional Optical-OFDM ANN-NLE case as shown in Equation (9), this is due to the fact that in Optical OFDM signals has real and imaginary components that the ANN-NLE has to deal with, where for the Fast-OFDM case the equalizer only deals with the real data [13].

$$FLOPS_{ANN} = Nop_{IL} + Nop_{HL} + Nop_{OL} = 2\left[N_{SC}{}^2 * M\right] + 2\left[(N_{SC} * M)^2 * (N_{HL} - 1)\right] + 2\left[N_{SC}{}^2 * M\right], \quad (9)$$

Below, Figure 8 illustrates a comparison between ANN-NLE and IVSTF-NLE in terms of subcarriers versus Floating points Operations (FLOPs). As the ANN-NLE is a modulation format dependent, therefore, different modulation formats M-ASK (M = 4, 8, and 16) has been used, moreover for the case of number of spans dependent IVSTF-NLE, the modulation formats has been set to 4 ASK, thereafter the number of span has been changed as follow 1 span, 10 spans, and 20 spans.

In terms of computational complexity FLOPS, it is shown that for all the number of subcarriers, ANN-NLE outperforms IVSTF-NLE. This is true even, when the best-case scenario of IVSTF-NLE, at 1 span, is compared to the worst-case scenario of ANN-NLE at 16 ASK, while fixing the number of subcarriers. This superiority is due to the utilization of a large number of low complexity neurons, which results in a low computational complexity when compared to the IVSTF-NLE that requires several FFT/IFFT blocks to perform the equalization.

It is worth noting that the proposed nonlinear equalizer can be very useful when applied in cost-sensitive metro-regional and short-reach networks. In future work, ANN-NLE will be implemented in wavelength division multiplexed (WDM) Fast-OFDM signals where IVSTF cannot tackle effectively inter-channel nonlinearities [30].

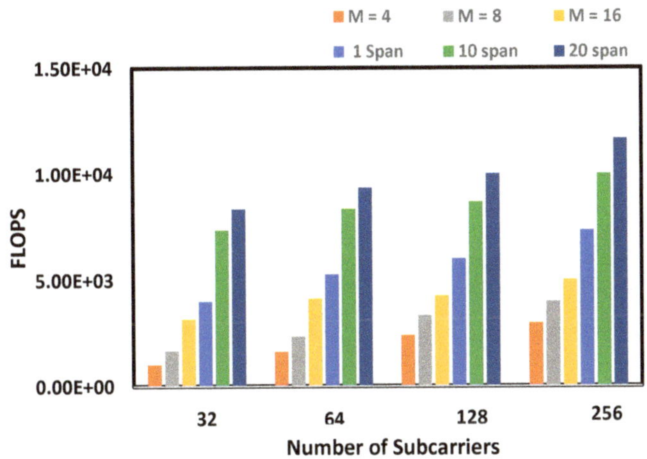

Figure 8. Computational complexity comparison between ANN-NLE and IVSTF-NLE versus the number of sub carriers, whereas the M-ASK modulation formats (M = 4, 8, and 16) corresponds to the ANN-NLE FLOPS and the 1 span, 10 span and 20 spans corresponds for the IVSTF-NLE FLOPS.

7. Conclusions

Reduction of intra-channel nonlinearities in low-cost direct-detected optical Fast-OFDM was numerically demonstrated, for the first time, using a low-complex version of an ANN nonlinear equalizer. In contrast to conventional ANN designs for coherent optical OFDM that employs two real-valued activation functions ("split" complex function) to process in-phase and quadrature components separately, here, only amplitude data were processed, thus significantly reducing the computational complexity of the equalizer (e.g., far less neurons were adopted). ANN-NLE proved to be a robust nonlinearity DSP technique for up to 19.37-Gb/s 16-ASKdirect-detected optical Fast-OFDM signals. It was shown that an 80 km extension in transmission-reach was feasible when using ANN compared to the benchmark Volterra-based NLE for 9.69-Gb/s 4-ASK modulated Fast-OFDM sub-carriers. This occurred because ANN can tackle stochastic nonlinear impairments such as parametric noise amplification. Finally, the dynamic parameters requirements of the sub-ranging quantizers were relaxed (DAC/ADC limitations) compared to linear equalization, namely the clipping ratio and quantization bits by 2 dB and 2 bits, respectively moreover 1 dB and 1 bit when compared with the IVSTF NLE performance. Regarding the computational complexity analysis, the ANN-NLE over-performs the back propagation equalizers such as the IVSTF due to the use of low complexity neurons where the IVSTF uses a large number of FFT/IFFT blocks as a part of the equalization process.

Finally, it is worth mentioning that the proposed nonlinear equalizer could be implemented in both state-art-of-the-art optical modems and future super-high-speed optical communication systems, supporting > 40-Gb/s per wavelength. More importantly, due to the low complexity of both Fast-OFDM based DSP and ANN-NLE, the proposed hybrid solution should be more practical for real-time signal processing than benchmark approaches (e.g., conventional optical OFDM and IVSTF or DBP).

Funding: This research received no external funding.

Acknowledgments: I also thank Elias Giacoumidis from Dublin City University for his valuable contributions on the simulation setup and editing.

Conflicts of Interest: The authors declare no conflict of interest.

References

1. Jansen, S.L.; Morita, I.; Schenk, C.W.; Takeda, N.; Tanaka, H. Coherent Optical 25.8-Gb/s OFDM Transmission over 4160-km SSMF. *IEEE J. Lightw. Technol.* **2008**, *26*, 6–15. [CrossRef]
2. Pratt, W.K. *Laser Communication Systems*, 1st ed.; John Wiley & Sons: New York, NY, USA, 1969.
3. Li, K.; Darwazeh, I. System performance comparison of Fast-OFDM with overlapping MC-DS-CDMA and MT-CDMA systems. In Proceedings of the 6th International Conference on Information, Communications & Signal Processing, Singapore, 10–13 December 2007; pp. 1–4.
4. Ibrahim, S.K.; Jian, Z.; Rafique, D.; O'Dowd, J.A.; Ellis, A.D. Demonstration of world-first experimental optical Fast OFDM system at 7.174 Gbit/s and 14.348 Gbit/s. In Proceedings of the European Conference and Exhibition on Optical Communication (ECOC), Torino, Italy, 19–23 September 2010; pp. 1–3.
5. Zhao, J.; Ibrahim, S.K.; Rafique, D.; Gunning, P.; Ellis, A.D. Symbol synchronization exploiting the symmetric property in optical Fast OFDM. *IEEE Photonics Technol. Lett.* **2011**, *23*, 594–596. [CrossRef]
6. Giacoumidis, E.; Tsokanos, A.; Mouchos, C.; Zardas, G.; Alves, C.; Wei, J.L.; Tang, J.M.; Gosset, C.; Jaouën, Y.; Tomkos, I. Extensive comparisons of optical Fast-OFDM and conventional optical OFDM for local and access networks. *J. Opt. Commun. Netw.* **2012**, *4*, 724–733. [CrossRef]
7. Jarajreh, M.A.; Ghassemlooy, Z.; Ng, W.P. Improving the chromatic dispersion tolerance in long-haul fibre links using the coherent optical orthogonal frequency division multiplexing. *IET Microw. Antennas Propag.* **2010**, *6*, 651–658. [CrossRef]
8. Gao, G.; Zhang, J.; Gu, W. Analytical Evaluation of Practical DBP-Based Intra-Channel Nonlinearity Compensators. *IEEE Photonics Technol. Lett.* **2013**, *25*, 717–720. [CrossRef]
9. Liu, L.; Li, L.; Huang, Y.; Cui, K.; Xiong, Q.; Hauske, F.N.; Xie, C.; Cai, Y. Intrachannel Nonlinearity Compensation by Inverse Volterra Series Transfer Function. *IEEE J. Lightw. Technol.* **2012**, *30*, 310–316. [CrossRef]
10. Lowery, A.J. Fiber nonlinearity pre- and post-compensation for long-haul optical links using OFDM. *Opt. Exp.* **2007**, *15*, 12965–12970. [CrossRef]
11. Giacoumidis, E.; Aldaya, I.; Jarajreh, M.A.; Tsokanos, A.; Le, S.T.; Farjady, F.; Ellis, A.D.; Doran, N.J. Volterra-based Reconfigurable Nonlinear Equalizer for Dual-Polarization Multi-Band Coherent OFDM. *IEEE Photonics Technol. Lett.* **2014**, *26*, 1383–1386. [CrossRef]
12. Giacoumidis, E.; Mhatli, S.; Nguyen, T.; Le, S.; Aldaya, I.; Ellis, A.; Mccarthy, M.; Eggleton, B. Comparison of DSP-based nonlinear equalizers for intra-channel nonlinearity compensation in coherent optical OFDM. *Opt. Lett.* **2016**, *41*, 2509–2512. [CrossRef]
13. Giacoumidis, E.; Lin, Y.; Wei, J.; Aldaya, I.; Tsokanos, A.; Barry, B. Harnessing machine learning for fiber-induced nonlinearity mitigation in long-haul coherent optical OFDM. *Future Internet* **2019**, *11*, 2. [CrossRef]
14. Jarajreh, M.A.; Giacoumidis, E.; Aldaya, I.; Le, S.T.; Tsokanos, A.; Ghassemlooy, Z.; Doran, N.J. Artificial neural network nonlinear equalizer for coherent optical OFDM. *IEEE Photonics Technol. Lett.* **2015**, *27*, 387–390. [CrossRef]
15. Zhao, J. Multi-Tap Equalization for Performance Improvement in Optical Fast OFDM Systems. *IEEE Photonics Technol. Lett.* **2015**, *27*, 42–45. [CrossRef]
16. Zidan, M.; Abdel-Aty, A.; El-shafei, M.; Feraig, M.; Al-Sbou, Y.; Eleuch, H.; Abdel-Aty, M. Quantum Classification Algorithm Based on Competitive Learning Neural Network and Entanglement Measure. *Appl. Sci.* **2019**, *9*, 1277. [CrossRef]
17. Walczak, S. An empirical analysis of data requirements for financial forecasting with neural networks. *J. Manag. Inf. Syst.* **2001**, *17*, 203–222.
18. Yue, F.; Chai, T. Neural-network-based nonlinear adaptive dynamical decoupling control. *IEEE Trans. Neural Netw.* **2007**, *18*, 921–925. [CrossRef] [PubMed]
19. Mellit, A.; Eleuch, H.; Benghanem, M.; Elaoun, C.; Massi Pavan, A. An adaptive model for predicting of global, direct and diffuse hourly solar irradiance. *Energy Convers. Manag.* **2010**, *51*, 771–782. [CrossRef]

20. Nsiri, B.; Mhatli, S.; Mallouki, N.; Ghanbarisabagh, M.; Giacoumidis, E.; Ammar, M.; Bouallègue, A.; Attia, R. Nonlinearity mitigation of optical Fast-OFDM signals using a Wiener-Hammerstein electrical equalizer. In Proceedings of the Science and Information Conference (SAI), London, UK, 28–30 July 2015; pp. 1056–1061.
21. Savazzi, P.; Favalli, L.; Costamagna, E.; Mecocci, A. A suboptimal approach to channel equalization based on the nearest neighbor rule. *IEEE J. Sel. Areas Commun.* **1998**, *16*, 1640–1648. [CrossRef]
22. Rajbhandari, S.; Ghassemlooy, Z.; Angelova, M. Effective denoising and adaptive equalization of indoor optical wireless channel with artificial light using the discrete wavelet transform and artificial neural network. *IEEE J. Lightw. Technol.* **2009**, *27*, 4493–4500. [CrossRef]
23. Charalabopoulos, G.; Stavroulakis, P.; Aghvami, A.H. A frequency-domain neural network equalizer for OFDM. In Proceedings of the GLOBECOM, San Francisco, CA, USA, 1–5 December 2003; pp. 571–575.
24. Giacoumidis, E.; Ibrahim, S.K.; Zhao, J.; Wei, J.L.; Tang, J.M.; Ellis, A.D.; Tomkos, I. Effect of ADC on the Performance of Optical Fast-OFDM in MMF/SMF-Based Links. In Proceedings of the Electromagnetics Research Symposium (PIERS), Suzhou, China, 12–16 September 2011; pp. 402–406.
25. Riedmiller, M.; Braun, H.A. Direct adaptive method for faster backpropagation learning: The RPROP Algorithm. In Proceedings of the IEEE International Conference on Neural Networks (ICNN), San Francisco, CA, USA, 28 March–1 April 1993; pp. 586–591.
26. Chen, E.; Tao, R.; Zhao, X. Channel equalization for OFDM system based on the BP Neural network. In Proceedings of the IEEE International Conference on Signal Processing (ICSP), Beijing, China, 16–20 November 2006.
27. Benvenuto, N.; Piazza, F. On the complex backpropagation algorithm. *IEEE Trans. Signal Proc.* **1992**, *40*, 967–969. [CrossRef]
28. Giacoumidis, E.; Wei, J.L.; Yang, X.L.; Tsokanos, A.; Tang, J.M. Adaptive modulation-enabled WDM impairment reduction in multi-channel optical OFDM transmission systems for next generation PONs. *IEEE Photonics* **2010**, *2*, 130–140. [CrossRef]
29. Giacoumidis, E.; Jarajreh, M.A.; Sygletos, S.; Le, S.T.; Tsokanos, A.; Hamié, A.; Pincemin, E.; Jaouën, Y.; Farjady, F.; Ellis, A.D.; et al. Dual-polarization multi-band OFDM transmission and transceiver limitations for up to 500 Gb/s in uncompensated long-haul links. *Opt. Exp.* **2014**, *22*, 10975–10986. [CrossRef]
30. Giacoumidis, E.; Le, S.T.; Aldaya, I.; Wei, J.L.; McCarthy, M.E., Doran, N.J.; Eggleton, B.J. Experimental Comparison of Artificial Neural Network and Volterra based Nonlinear Equalization for CO-OFDM. In Proceedings of the OSA Optical Fiber Communication Conference and Exposition and the National Fiber Optic Engineers Conference (OFC/NFOEC), San Francisco, CA, USA, 20–24 March 2016.

applied sciences

MDPI

Review

AI-Based Modeling and Monitoring Techniques for Future Intelligent Elastic Optical Networks

Xiaomin Liu, Huazhi Lun, Mengfan Fu, Yunyun Fan, Lilin Yi, Weisheng Hu and Qunbi Zhuge *

Shanghai Institute for Advanced Communication and Data Science, State Key Laboratory of Advanced Optical Communication Systems and Networks, Department of Electronic Engineering, Shanghai Jiao Tong University, Shanghai 200240, China; xiaomin.liu@sjtu.edu.cn (X.L.); huazhi.lun.sjtu@gmail.com (H.L.); mengfan.fu@sjtu.edu.cn (M.F.); fanyunyun@sjtu.edu.cn (Y.F.); lilinyi@sjtu.edu.cn (L.Y.); wsh@sjtu.edu.cn (W.H.)

* Correspondence: qunbi.zhuge@sjtu.edu.cn

Received: 27 November 2019; Accepted: 27 December 2019; Published: 3 January 2020

Abstract: With the development of 5G technology, high definition video and internet of things, the capacity demand for optical networks has been increasing dramatically. To fulfill the capacity demand, low-margin optical network is attracting attentions. Therefore, planning tools with higher accuracy are needed and accurate models for quality of transmission (QoT) and impairments are the key elements to achieve this. Moreover, since the margin is low, maintaining the reliability of the optical network is also essential and optical performance monitoring (OPM) is desired. With OPM, controllers can adapt the configuration of the physical layer and detect anomalies. However, considering the heterogeneity of the modern optical network, it is difficult to build such accurate modeling and monitoring tools using traditional analytical methods. Fortunately, data-driven artificial intelligence (AI) provides a promising path. In this paper, we firstly discuss the requirements for adopting AI approaches in optical networks. Then, we review various recent progress of AI-based QoT/impairments modeling and monitoring schemes. We categorize these proposed methods by their functions and summarize advantages and challenges of adopting AI methods for these tasks. We discuss the problems remained for deploying AI-based methods to a practical system and present some possible directions for future investigation.

Keywords: optical transmission; optical networks; machine learning; artificial intelligence; quality of transmission; optical performance monitoring; failure management

1. Introduction

The progress of 5G mobile networks, internet of things and cloud services has raised high demands and new requirements for the capacity and reliability of optical networks. To serve the rapidly increasing number of internet service users, the technologies of optical networks are continuously evolving. The development of elastic optical networks (EON) [1] enables network controllers to scale up or down resources in order to utilize spectrum resources efficiently [2]. However, the EON architecture increases network complexity because of the various configurations of links and signals, which makes it more challenging to maintain the high transmission quality of a lightpath from the beginning of life (BoL) to the end of life (EoL). Since a large amount of data is transmitted in each link, even a brief disruption of traffic flows can lead to disastrous degradation [2]. Therefore, improving the reliability of optical networks is also important.

To reach a high capacity, optical networks should better utilize network resources. In many scenarios, since a planning tool cannot accurately estimate the quality of transmissions (QoT), a high design margin is mandatory, which accounts for the difference between the planned metrics and the real value to ensure proper operations of networks [3]. A high margin can lead to the underutilization of spectrum resources. Therefore, to build a low margin optical network to increase network capacity, a more accurate planning tool is needed to estimate the QoT prior to link deployment or reconfiguration [4]. In this case, an accurate QoT model is essential and impairment models can improve the accuracy of the QoT model. On the other hand, to improve the reliability of optical networks, controllers should be capable of obtaining the real-time status of networks to prevent the serious degradation of systems. To achieve this, advanced optical performance monitoring (OPM) techniques are essential to enable needed functionalities to monitor the QoT and impairments. If failures occur in optical networks, the monitoring mechanisms should be capable of detecting, identifying and localizing them. Therefore, in summary, the modeling and monitoring techniques are the key building blocks for the next generation EON. The basic architecture of the modeling and monitoring techniques is shown in Figure 1.

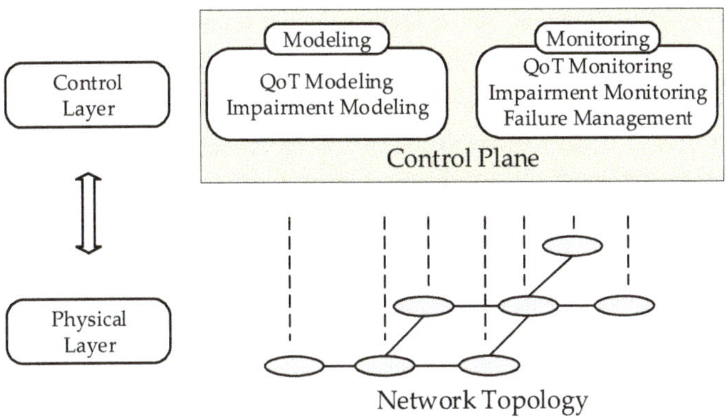

Figure 1. Applications of modeling and monitoring techniques in optical networks.

For the modeling, some models are applied to judge whether one lightpath meets the requirement for establishment in terms of the QoT [4]. Some are applied to estimate the specific value of the QoT or impairments [5]. In EON, there are some challenges for traditional analytical models. Firstly, there exists typically a tradeoff between complexity and accuracy. Some sophisticated analytical models, e.g., the split-step Fourier method (SSFM) [6], are capable of capturing different impairments with great precision, but the complexity may be prohibitively high. Some approximate models, e.g., the Gaussian noise (GN) model [7], can be calculated in a short time, but the accuracy needs to be improved especially for heterogeneous and dynamic links. Moreover, because of the diversity of EON, it is difficult to obtain one specific model for all scenarios. In this case, the estimation results of models may appear a large deviation for some scenarios.

Artificial intelligence (AI) [8] technologies provide new opportunities to solve these problems. In many scenarios, machine learning (ML) methods can obtain a higher accuracy and/or a lower complexity compared to analytical models. For instance, in [5], an artificial neural network (ANN) is adopted to estimate fiber nonlinear noise more accurately and efficiently compared to the original analytical model. The accuracy of this ANN-based nonlinear estimator is higher than the incoherent GN (IGN) model and the complexity is much lower than the SSFM. Moreover, for situations where there is no suitable traditional model, ML methods can make estimations utilizing the data extracted from simulations or real scenes. For example, the filtering effect brought by reconfigurable optical add-drop multiplexers (ROADM) can be modeled with an ANN [9]. Finally, many data-driven methods with ML can be adopted to adjust analytical models to be scalable for more scenarios where they show large deviations. For instance, in [10], ML algorithms are used to improve the performance of the analytical model with data collected from an established lightpath.

The transmission performance of an established light path is not always reliable due to the various changes of link conditions. Therefore, optical performance monitoring (OPM) is a key building block, which enables network controllers to adjust link configurations according to the real-time status of a system. Moreover, monitoring results can be used to detect, identify and localize failures in EON's. However, the heterogeneity of EON's has also raised many new challenging requirements for the monitoring techniques, and ML shows a potential in building more intelligent and efficient monitoring schemes. Firstly, faster response time is desired for monitoring [2]. Since a monitoring agent should provide information for optimizing lightpath configurations and diagnosing the anomaly, the monitoring scheme needs to be capable of tracking the change of the network performance. According to [2], the monitoring time of some network applications is required to be at the order of milliseconds. Therefore, some traditional methods with complex data processing and a long-time window may not be compatible with dynamic real-time applications. To solve this problem, advanced ML methods with forward propagation mechanisms [11], such as ANN, convolutional neural networks (CNN) and so on, can be employed to accomplish the feature extraction and estimate real-time status in a short time period [5,12,13]. These monitoring tools can be trained offline before deployment. When estimating the signal performance, the pre-trained monitors can respond in a very short time. Secondly, monitoring techniques should be cost-effective [2]. In particular, they should not necessitate expensive external devices, and one OPM block is preferable to monitor multiple impairments. It may be difficult for analytical models to achieve these two goals simultaneously but ML-aided methods can help to fulfill these requirements. For instance, samples of received signals can be input to ML algorithms for monitoring the chromatic dispersion (CD), polarization-mode dispersion (PMD) and optical signal-to-noise ratio (OSNR) at the same time [14]. Moreover, when obtaining information from the receiver digital signal processing (DSP) modules, ML methods may be able to monitor the QoT or impairments without any external devices such as the optical spectrum analyzer (OSA) [15].

Therefore, for the next generation EON, applications of ML techniques for modeling and monitoring can provide strong support to build a reliable and intelligent optical network with lower design margins. This paper is intended to review recent progress in AI-enabled modeling and monitoring techniques for EON. Since optical networks are full of data with heterogeneous sources and various characteristics, it is possible to improve the accuracy and/or sensitivity of optical performance estimation functionalities with these data. However, the large number of data also makes it more challenging to discover useful information from them. In this case, data-driven ML methods are essential tools for network planning and management, but these methods should be improved to be cost-effective and reliable for deployment. Several previous review works have provided comprehensive summaries of the applications of ML techniques in optical networks [2,16–19]. They discuss the ML-based techniques adopted in various domains and point out many possible directions for the future deployment strategies. In this paper, we focus on the AI-based techniques specifically for link modeling and monitoring in optical networks. In addition, we discuss and summarize the advantages and challenges for adopting the AI-based modeling and monitoring methods in the future EON. This paper is organized as follows.

- In Section 2, we firstly introduce the background and challenges for modeling the QoT and impairments in EON's. The potentials of applying ML to estimate network performance are also discussed. Then, we review many previous works on ML-based modeling techniques.
- In Section 3, we firstly review various previous works on ML-based monitoring techniques. Afterwards, the monitoring techniques specifically for failure management are elaborated.
- In Section 4, the use cases for AI-based modeling and monitoring techniques are discussed.
- In Section 5, we provide a lookout for the future of utilizing ML methods in EON by discussing both the challenges and opportunities.
- In Section 6, a conclusion for this paper is provided.

2. AI-Based QoT and Impairment Modeling

2.1. Background and Challenges

QoT modeling for an unestablished lightpath can help planning tools in the control plane to develop proper strategies of routing, wavelength assignment and signal configurations [20–25]. In EON, during the phase of network planning, the accuracy of QoT and impairment models is influenced by various configurable parameters like modulation format, symbol rate and physical path in optical networks. If these parameters are not accurate, the estimations of QoT may have deviations compared with the real value [5,26,27]. In this case, due to the inaccuracy of planning tools, a large design margin [3] is needed and networks are overutilized to avoid network degradation until the EoL. As a result, QoT models with a higher accuracy are desired and impairment models can provide an insight into the contributions of each individual impairment to help QoT estimators reach a better performance.

For the QoT modeling, some traditional methods [28] can estimate the performance of an optical link in terms of signal-to-noise ratio (SNR), pre-forward error correction (FEC) bit error rate (BER), OSNR and so forth. For the impairment modeling, traditional methods can estimate some important physical layer effects, such as fiber nonlinearity, optical filtering effect and amplified spontaneous emission (ASE) noise. The requirements for QoT and impairment modeling techniques of the next-generation EON are illustrated as follows.

- Self-adaptiveness: Analytical models are essential for estimating the QoT of unestablished lightpaths. However, they may not be scalable for all scenarios since the assumptions for these models may be inappropriate when the configuration of traffic optical paths evolves continuously. For instance, the optical amplifier gain spectrum is wavelength-dependent but some models assume the gain to be identical for all channels. This kind of improper assumption may lead to an inaccurate estimation of the ASE noise. Therefore, network planning tools with self-adaptive QoT and impairment models are highly desired to guarantee a high-quality transmission from the BoL to the EoL.
- Efficiency: For many QoT and impairment models, traditional models with high precision may incur burdensome computational requirements. For example, to model the nonlinear impairment, the SSFM [6,29–31] can reach a high accuracy if the step size is sufficiently small, which leads to a high complexity. The GN model [7] can provide results in a very short time but the precision is lower than that of the SSFM in most scenarios. Therefore, models that can efficiently make estimations with a high accuracy are desired.
- High tolerance to parameter uncertainty: In a practical system, link parameters can be uncertain due to inaccurate measurements and other reasons. If the uncertainty of the model input exists, there might be a significant deviation between the real value and the model estimation [5]. Therefore, models that are less sensitive to parameter uncertainty are also desired.

To fulfill these requirements, data-driven ML methods open new opportunities. Firstly, ML methods are mostly data-driven [32], which means they enable the model to learn the characteristics of the dataset, in principle even without any theoretical information [4,33–36]. This specific ability of learning adaptively with data allows ML models to be easily extended to any scenarios if the simulation, experiment or field-trial data for this situation can be obtained [13,23,37]. Secondly, for most optical networks, the number of tunable parameters for link configurations is limited. Therefore, the number of input parameters for QoT or impairment models are relatively small [5,33,38], which enables ML models to reach a good performance with simple structures such as the ANN with a small number of nodes and hidden layers [23]. In this case, these low-complexity ML models can calculate faster compared with some traditional models. Many previous works using a simple ANN or linear regression have already achieved good performances [5]. Finally, advanced ML algorithms like ensemble learning [39] and Theil-Sen regression [40] can address the drawbacks of the least squared algorithms and make models less sensitive to the outliers and fluctuations of data. Besides, training techniques like data augmentation [41,42] can improve the model robustness to parameter uncertainty and avoid overfitting by adding interference manually. In this section, we reviewed various previous works for AI-based QoT and impairment modeling techniques.

2.2. AI-Based QoT Modeling

For the QoT modeling, there are many types of metrics, such as BER, Q-factor, SNR, OSNR and margin. The aim of the QoT modeling is to precisely estimate the link performance and then build low margin networks. The requirement of the QoT estimations differs in different scenarios. Some need to judge whether one light path can be established or not [4,21,38], and some need the specific value of the QoT metrics. For the former, ML classification methods [43] can be used such as K-nearest neighbors (KNN), random forests (RF), support vector machine (SVM), logistic regression (LR), ANN and so forth. For the latter, ML regression methods [43,44] can be employed such as network Kriging (NK), Gaussian process (GP), CNN [45], ANN and so forth. We provided a review of some recent ML-based QoT modeling techniques in the literature for different metrics in this section. They are listed in Table 1 and elaborated as follows.

Table 1. Summary of the machine learning (ML)-based quality of transmission (QoT) modeling techniques discussed in Section 2.2.

Modeling Targets	Algorithms	Input Features
BER	K-Nearest Neighbors, Random Forest [4]	Traffic volume, modulation format, total length of links, length of the longest link, number of lightpath links
	Stochastic Gradient Descent Polynomial Regression [21]	Generalized OSNR, baud rate, modulation format, FEC, slot-size
	Deep Graph Convolutional Neural Networks [46]	Total length of the path, span length, central frequency, number of slots in each path, modulation format, number of Erbium-doped fiber amplifier (EDFA), number of links, BER
Q-factor	Case-based Reasoning [47,48]	Route, selected wavelength, total length of the path, sum of the co-propagating lightpaths per link, standard deviation of the number of total co-propagating lightpaths
	Transfer Learning [33]	Channel loading, per-channel output power
OSNR	Network Kriging, l_2-norm Regularization [49]	Average PMD of each link, accumulation value of CD, the self-phase modulation (SPM) quantified through the nonlinear phase of the signal
	Gaussian Process Regression [50]	Wavelength, OSNR of the established wavelength
SNR	Combination of Machine Learning and Physical Layer Model [10]	Lightpath length, link load, number of crossed EDFAs
	Gradient Decent [51]	Power, noise figure
Margin	K-Nearest Neighbors, Linear Regression, Support Vector Machine, Artificial Neural Networks [38]	Number of hops, number of spans, total link length, average link length, maximum link length, average span attenuation, average CD

For the BER estimation, in [4], an ML-based classifier is used to decide whether the BER of an unestablished lightpath can achieve the network requirement. Features of the model are the traffic volume, modulation format, lightpath total length, length of the longest link and number of lightpath links. The training dataset is obtained from the deployed lightpaths. The employed ML classifier algorithms are KNN and RF with various kernel settings. Moreover, this work comprehensively compare the performance of different ML algorithms. The influences of different combinations of input features and different sizes of dataset are also analyzed. The result shows that RF outperformes KNN in accuracy and efficiency in most cases. The result also shows that a bigger dataset can help to reach a higher accuracy. In [21], the generalized optical signal-to-noise ratio (gOSNR), baud rate, modulation format, FEC, slot-size and so on are used to estimate the BER and the training data is obtained from a practical system. Therefore, this model can enable controllers to find the optimum configuration of a light path for each specific network. In [46], a deep graph convolutional neural network (DGCNN) is applied to estimate the feasibility of the network state. This work considers the crosstalk between unestablished and established light paths according to historical data.

For the Q-factor estimation, in [47,48], a cognitive QoT estimator classifies lightpaths to high- or low-quality categories. The classification method is case-based reasoning (CBR), which is based on the prior experiences or cases to make estimations. Features for this model include the route, selected wavelength, total length of a path, sum of the co-propagating lightpaths per link and standard deviation of the number of total co-propagating lightpaths. To extend a pre-trained model to more scenarios, transfer learning is proposed in [52] to make use of collected data from new scenes for retraining. This method can effectively reduce the training time when configurations of the optical networks change. Methods mentioned above all use historical data from real scenes and they all achieve a good performance for estimating the QoT. Therefore, we can infer that data-driven ML methods can improve the training efficiency and the scalability of models to more systems.

For the OSNR estimation, in [49], regression methods like network Kriging (NK) and least-squares minimization with l_2-norm regularization are utilized. The parameters used for estimations are the average PMD of each link, accumulation value of CD, and SPM quantified through the nonlinear phase of the signal. The algorithm is based on established light paths to evaluate an unestablished path for transparent optical networks. This method successfully helps to design a reliable light path efficiently. According to [50], in some practical systems, the noise figure and gain of amplifiers and fiber loss are wavelength-dependent. In this case, the Gaussian process regression (GPR) is used to estimate the OSNR with a confidence output.

For the SNR estimation, in [10], the combination of the ML model (ML-M) and physical layer model (PLM) is applied to build a framework called ML-PLM to estimate the QoT performance. This model is based on the data from the existing connections of a network. Features used for estimation are the light path length, link load and number of crossed Erbium-doped fiber amplifiers (EDFAs). The simulation shows that, this method can reduce the influence of the uncertainty of parameters such as the fiber attenuation, dispersion, nonlinear coefficients or amplifier noise. Moreover, the more light paths the model can get from the network topology, the higher accuracy the model can achieve. In this way, ML-PLM can reach better performances and makes the model suitable for a dynamic network. In [51], gradient decent is used to correct the deviations of the input parameters for the QoT estimators. This method takes advantage of back-propagation algorithms embedded in many neural networks, which successfully reduces the uncertainty of models.

For the margin estimation, in [38], ML models such as KNN, LR, SVM and ANN are proposed to judge whether the residual margin is positive. The input features are the number of hops, number of spans, total link length, average link length, maximum link length, average span attenuation and average dispersion. To build a better classifier, those models for classification are investigated with different kinds of kernels. Then, to obtain the specific value of the residual margin an ANN is employed. In [38], the performances of the adopted ML algorithms are compared with each other and they all reach a decent performance.

2.3. *AI-Based Impairment Modeling*

Accurate modeling of impairments can provide more information to improve the accuracy of QoT models. Moreover, the estimation of specific impairments can help controllers design an optimum configuration of a light path. In this section, since impairments like CD and PMD can be compensated in the receiver, we focus on the impairments that may cause performance degradation. A few recent works using AI-based modeling methods for estimating fiber nonlinearity, filtering effect and ASE noise are investigated in this section. They are listed in Table 2 and introduced as follows.

Table 2. Summary of the ML-based impairment modeling techniques discussed in Section 2.3.

Modeling Targets	Algorithms	Input Features
Nonlinear Noise	Artificial Neural Networks [5]	Nonlinear SNR from the GN model, span number, maximum span length, average span length, optical launch power, link length, net CD, average gamma of fiber spans, average attenuation of fiber spans, number of wavelength-division multiplexing (WDM) channels
Optical Filtering Effect	Artificial Neural Networks [9]	Number of ROADMs, OSNR, loaded noise distribution, bandwidth distribution
Gain Spectrum of EDFA	Deep Neural Networks [34]	Power levels of all WDM channels
EDFA Gain Excursion	Multilayer Perceptron [35]	Gain setting, total input power, input power of each channel

2.3.1. Nonlinear Noise

For the nonlinear effect modeling, sophisticated analytical models such as the SSFM [6] can provide accurate estimations. However, these methods also result in a long computation time. Although approximate models can calculate much more quickly [53], they cannot guarantee the accuracy in all scenarios, thus leading to a high design margin and an inefficient utilization of network resources [3]. In [5], a combination of analytical models and ML methods is proposed to reach a higher accuracy for nonlinear noise estimation.

2.3.2. Filtering Effect

In future EON, ROADM can enable optical networks to support the flexible multiplexing and demultiplexing, which is important for build an intelligent network with more capacity and dynamicity. However, in this case, the filtering effect caused by cascaded ROADMs can also influence the QoT much more significantly because of the reduced guard band between channels. In [9], an ANN-aided approach is introduced to estimate the filtering effect. The input features of the neural network are the ROADM number, OSNR, loaded noise distribution and bandwidth distribution. A one-hidden layer ANN can estimate the SNR of the light path induced by the filtering effect with error mostly less than 1 dB. In practical systems, the filtering effect can be more significant when multiple impairments co-exist such as nonlinearity. Besides, the filtering effect is not a kind of additive noise and SNR may not be the best metric for evaluation. Therefore, problems like how to model the filtering effect together with other impairments and how to quantify the filtering effect using a proper metric should be further investigated.

2.3.3. ASE Noise

In a practical system, to accurately model the ASE noise generated by EDFAs, the noise figure (NF) of each EDFA at each wavelength should be precisely known. According to [36], the NF of an EDFA is related to the gain at each wavelength. Therefore, the ASE noise can be more accurately estimated with the aid of an accurate EDFA gain model. However, the spectral hole burning [54] (SHB) effect makes the spectral gain profile of an EDFA change dynamically under channel reconfigurations, thus leading to a power excursion. Since it is hard for the traditional model to efficiently model the gain spectrum of an EDFA with different power loadings in each channel, data-driven ML methods can be adopted. In [34], deep learning is adopted to estimate the gain of each channel individually. To simplify the structure of ML algorithms, a multilayer perceptron neural network is introduced to estimate the gain of all channels at the same time [35].

3. AI-Based Optical Performance Monitoring

OPM is key to ensure the reliability of optical networks [16]. According to [2], monitoring techniques can enable several essential and advanced network functionalities. Firstly, a precise monitoring of QoT and impairments [55,56] can make the control plane accurately assess the signal quality. Therefore, the monitoring information can guide the network self-reconfiguration and also enables receivers to adapt some impairment compensation algorithms. Secondly, the real-time monitoring can continuously obtain the condition of the physical layer. If the QoT deteriorates, monitoring agents can detect failures. Then, the controller can reconfigure the network to avoid further degradations. Finally, monitoring data from real scenes can be used to retrain the planning model. This retraining scheme can improve the accuracy of planning tools and make the design margin lower. At the same time, there are also some challenging requirements for deploying an OPM in an EON, such as how to track the real-time change of the optical networks accurately in a short response time and how to monitor multiple impairments simultaneously. These challenges have been elaborated in Section 1. ML shows its potential to fulfill these challenges. In this section, we review various works using ML for OPM. According to their different functions, these approaches are divided into two categories. We firstly introduce some use cases of monitoring the QoT and impairments of a lightpath. Then, we review the monitoring techniques for detecting, identifying and localizing soft failures in a network. These two aspects are discussed as follows.

3.1. AI-Based QoT and Impairment Monitoring

For the QoT monitoring, the evaluation of BER, SNR, Q-factor, QSNR and so forth can enable controllers to assess the transmission performance of each established light path and provide a quantitative measure to check whether the designed QoT can be ensured. At the same time, impairment monitoring is also needed to provide an insight into each specific effect in the physical layer. In this section, various applications of ML for monitoring QoT and impairment are discussed. A brief summary of methods discussed in this section is shown in Table 3 and the details are elaborated as follows.

In [14], an ANN is used to monitor the OSNR, CD and PMD simultaneously with empirical asynchronously sampled signal amplitudes. In [57], to make an easier monitoring procedure without labor-intensive feature engineering, deep neural networks (DNN) are used to monitor the OSNR with asynchronously sampled raw data. For this work, neural networks with an advanced structure perform the feature extraction and monitoring calculation at the same time. Moreover, the results show that a larger training dataset and a deeper neural network can help to increase the estimation performance. As more advanced neural network structures emerge, CNN is also introduced to monitor the OSNR and modulation format simultaneously [13,58,59]. In [37], ANN is adopted to monitor the OSNR based on the historical data collected from real systems. In [60], principle component analysis (PCA) and ANN are used to monitor the OSNR, bit rate, modulation format, CD and DGD by asynchronous delay-tap plots. In this case, PCA can reduce the number of input parameters, thus reducing the complexity of the ANN. A similar approach is investigated in [61] to monitor the OSNR and identify the modulation format by asynchronous single channel sampling, which makes the algorithms simple and low-cost. In some other situations, ML-methods are also employed to monitor specific impairments. In [62], DNN is proposed to monitor the OSNR and modulation format with signals' amplitude histograms. This method only requires few DSP blocks, which makes it cost-effective for deployment. In [63], kernel-based ridge regression is used to monitor the CD and differential group delay (DGD) simultaneously. This method is validated by simulations and experiments. In [64], the long short-term memory (LSTM) neural network is applied to monitor the OSNR with the four-tributary digital outputs. The mean absolute error can be significantly reduced from 0.4 to 0.04 dB compared with other

ML algorithms. In [65], OSNR and nonlinear noise power are monitored simultaneously based on frequency domain signals. In [66], to identify the impairment causing the transmission degradation, SVM can accurately make classifications between CD, PMD and noncoherent crosstalk.

Table 3. Summary of the ML-based monitoring techniques discussed in Section 3.1.

Algorithms	Features	Monitoring Targets
Artificial Neural Networks [14]	Empirical asynchronously sampled signal amplitudes	OSNR, CD, PMD
Deep Neural Networks [57]	Asynchronously sampled raw data	OSNR
Convolutional Neural Network [58]	Constellation-diagram	OSNR, Modulation Format
Convolutional Neural Networks [13,59]	Horizontal and vertical polarization, in-phase and quadrature-phase components of optical signals	OSNR, Modulation Format
Artificial Neural Networks [37]	Launched power, EDFAs' input powers, EDFAs' output powers, EDFAs' gains, EDFAs' NFs, etc.	OSNR
Principal Component Analysis, Artificial Neural Networks [60]	Asynchronous delay-tap plots	OSNR, CD, Differential Group Delay (DGD), Joint Bit-rate and Modulation Format Identification (BR-MFI)
Principle Component Analysis, Artificial Neural Networks [61]	Asynchronous single channel sampling data	OSNR, Modulation Format
Deep Neural Networks [62]	Signals' amplitude histograms	OSNR, Modulation Format
Kernel-based Ridge Regression [63]	Asynchronous delay-tap sampling data	CD, DGD
Long Short-Term Memory Neural Networks [64]	Four-tributary digital output	OSNR
Long Short-Term Memory Neural Networks [65]	Frequency domain signal	OSNR, Nonlinear Noise Power
Support Vector Machine [66]	Eye diagrams	CD, PMD, Noncoherent Crosstalk
Artificial Neural Networks [67]	Amplitude noise and phase noise correlation (ANC, PNC), number of WDM channel, total CD	Nonlinear SNR
Artificial Neural Networks [68]	Accumulative logarithmic ANC (ALANC), number of WDM channel, total CD, noise tangential and normal component	Nonlinear SNR
Support Vector Regression [69]	Amplitude noise correlation, logarithmic accumulated CD	Nonlinear SNR
Artificial Neural Networks [5]	Nonlinear SNR from GN model, span number, maximum span length, average span length, the launch power, link length, net CD, average gamma of fiber spans, average attenuation of fiber spans, number of WDM channels, ANC, PNC	Nonlinear SNR

In many scenes, obtaining specific features strongly related to an impairment can improve monitoring accuracy. In [68], the amplitude noise correlation (ANC) and phase noise correlation (PNC) are proved to be related to nonlinear impairments and an ANN is applied to monitor the nonlinear SNR based on them. In [69], multiple logarithmic ANCs are directly input to an estimator using support vector regression for monitoring the nonlinear SNR, which can estimate nonlinear noise without features like the number of WDM channels. Moreover, in [5], the ANC and PNC are combined with an analytical model such as the GN model to estimate nonlinear noise. Simulation results in [5] prove that this combination can improve the monitoring accuracy.

3.2. AI-Based Failure Management

Link failures can be classified into hard failures and soft failures. Hard failures in the link cause immediate disruptions but can be easily detected and restored. Soft failures just gradually deteriorate the performance of the link and they are hard to be detected. In addition, the causes behind them are challenging to be identified. Therefore, detecting and identifying soft failures are of great importance and highly desired. In this section, we review some recent works for failure management based on AI techniques and they are listed in Table 4.

Table 4. Summary of the AI-based failure management techniques discussed in Section 3.2.

Targets	Algorithms	Input Features
Detection, Identification	Finite State Machine [70]	BER, received power in receiver
Detection, Identification	Random Forests, Support Vector Machine [71]	The trend of BER
Identification	Convolutional Neural Networks [12]	Optical spectrum
Detection	Support Vector Machine, Decision Trees [15]	Optical spectrum
Detection	One-class Support Vector Machine [72]	Tap value of the adaptive filter

For the soft failure detection, current detection methods in a deployed network usually rely on a pre-defined threshold. However, because of the high complexity of modern optical networks, it is hard to set an accurate threshold. If it is set too loose, some soft failures may be ignored, and if too tight, false detection may occur. For soft failure identification, it is generally difficult to accomplish accurate identifications using analytic methods. To address the challenges faced by the traditional methods, many works are proposed to utilize the ML techniques to perform failure detection and identification. In [70], finite state machine (FSM) is used to detect and identify the soft failures caused by laser and wavelength selective switch (WSS). In [71], the trend of the BER is monitored and analyzed. The statistical characteristics of BER are input to the RF and SVM to detect the soft failure, and an ANN with a hidden layer is applied to identify the cause of the soft failure between EDFA and WSS. In [15], the optical spectrum is monitored using an optical spectrum analyzer (OSA). The features of it are extracted and analyzed to detect the soft failure caused by WSS. Then, controllers identify the anomaly between filter shift (FS) and filter tightening (FT). In [72], the tap value of the adaptive filter is analyzed using one-class SVM to detect the soft failure caused by laser, WSS and fiber nonlinearity. To summarize, ML techniques pave a promising way to address the problems of failure detection and identification. With the powerful learning capability of ML, the hidden patterns of the monitored data can be learned to enable various failure management functionalities. As optical networks becoming more dynamic and heterogeneous, traditional techniques for soft failure detection and identification may not be able to adapt to the complex scenarios well. Therefore, more applications of ML techniques are expected to be investigated in this field.

4. Use Cases

4.1. Use Case 1: AI-Based Nonlinear Noise Modeling

A use case for modeling the nonlinear SNR with ML is discussed as below. This use case is based on the methods proposed in [5]. The structure of the ML-based estimator is shown in Figure 2a. For this model, an analytical model provides a relatively low-accuracy result in a short time. Afterwards, the pre-calculated result is input to a ML engine together with the processed system features related to nonlinear interference. The system features are shown in Table 5. These features can be easily obtained by a central controller and the processing time is short. For this modeling scheme, the GN model can provide an approximate value with lower precision compared with the SSFM, and the ANN only needs to learn the residuals between the real value and the approximate one. In this way, only with a simple-structure ANN, the estimation result can be accurate. The simulation setup is shown in Figure 2c and the detailed description can be found in [5]. In Figure 2b, results show that when combining the ANN with the coherent GN model (CGN) or the IGN model, the estimation accuracy can be significantly improved.

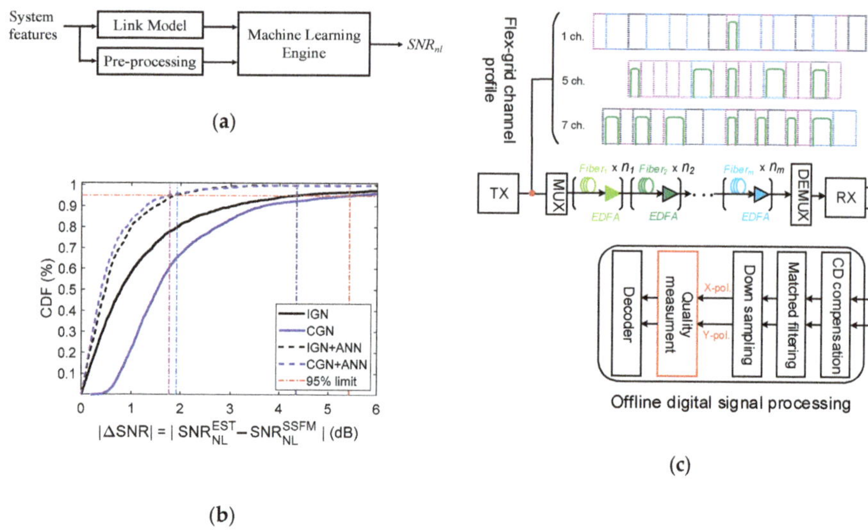

Figure 2. (**a**) The structure of the proposed modeling scheme in [5]. (**b**) The comparison of cumulative distribution functions (CDF) [5]. SNR_{NL}^{SSFM} means the nonlinear SNR estimated by the SSFM. SNR_{NL}^{EST} means the estimation made by the model proposed in [5]. ΔSNR means the estimation error between the SNR_{NL}^{EST} and the SNR_{NL}^{SSFM}. (**c**) The simulation setup.

Table 5. Summary of the modeling input features used in [5].

Features of modeling	1. SNR_{NL} from the GN model
	2. Span number
	3. Maximum span length
	4. Average span length
	5. Launch power
	6. Link length
	7. Net CD
	8. Average gamma of fiber spans
	9. Average alpha of fiber spans
	10. Number of WDM channels

4.2. Use Case 2: AI-Based Nonlinear Noise Monitoring

As elaborated in Section 3.1, many ML-based methods are proposed to monitor the nonlinear SNR in [66,67,72]. To improve the monitoring accuracy, in [5], the AI-based monitoring method combines the analytical models and the monitoring features such as ANC and PNC. As shown in Figure 3a–c, when combining monitoring features with analytical models, the maximum error reduces from 1.2 to 1 dB and 0.8 dB using the IGN and CGN, respectively. Moreover, the comparison of the CDF in Figure 3d also shows that the CGN model outperforms the IGN model to improve the ANN performance by 0.35 dB. In this work, the analytical model provides an approximate estimation. Afterwards, monitoring features are applied to improve the estimation accuracy based on the prior approximate estimations made by analytical models. Therefore, we can infer that ML can reach a higher accuracy if the input features are selected and processed properly.

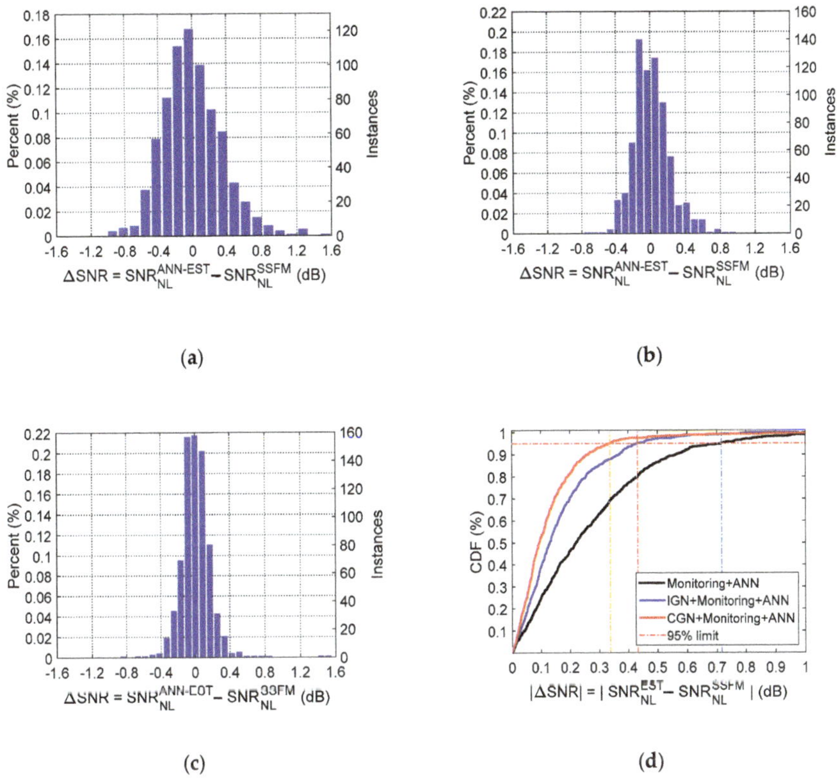

Figure 3. The estimation performance of the monitoring scheme proposed in [5]. (**a**) The error histogram of monitoring performance without any analytical model input. (**b**) The error histogram of monitoring performance with the incoherent Gaussian noise model (IGN). (**c**) The error histogram of monitoring performance with the coherent GN model (CGN). (**d**) The CDF of three monitoring strategies proposed in [5]. $SNR_{NL}^{ANN-EST}$ and SNR_{NL}^{EST} means the nonlinear signal-to-noise ratio (SNR) estimated by methods proposed in [5]. SNR_{NL}^{SSFM} means the nonlinear SNR estimated by the SSFM. ΔSNR means the estimation difference between the proposed method and the SSFM.

4.3. Use Case 3: AI-Based Soft Failure Identification

A use case for failure identification is elaborated in [12]. In addition to the filtering effect of WSS and ASE noise, fiber nonlinearity is also considered. Compared with the previous works, a deep learning algorithm is used and the power spectrum density (PSD) is extracted from a coherent receiver. The overall architecture is shown in Figure 4.

The SDN agent monitors the physical layer continuously and uploads the PSD to the control layer. Once the anomaly is detected, the CNN embedded in the anomaly identification module analyzes the PSD stored in the database. Finally, the identification results are output to the failure management module and proper actions are taken to restore the optical link.

The identification results are shown in Figure 5a. The results demonstrate a high accuracy of the proposed method when there exists only one type of anomaly. In the scene when multiple types of anomalies exit, the probability output by the SoftMax layer is utilized to gain insight into their respective influences on the system. The result is shown in Figure 5b. The influences of ASE and nonlinear interference (NLI) on the system are similar at first since the output probabilities of the two causes are both about 50 percent. Then, with the OSNR increasing, the NLI gradually becomes the dominant cause.

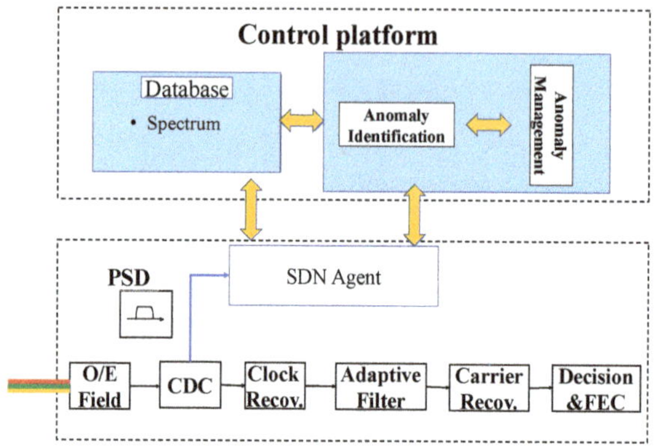

Figure 4. The overall architecture of the failure identification scheme in [12].

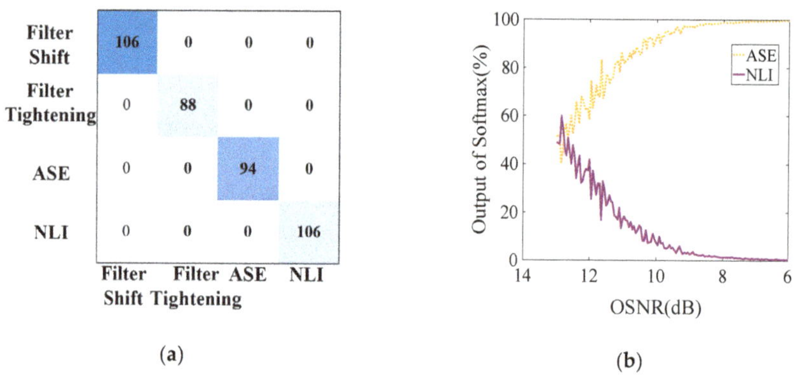

Figure 5. The performance of the method in [12]. (**a**) The accuracy of the proposed method. (**b**) The probability information output by the softmax layer.

5. Future Work

To build a reliable optical network with a lower margin, ML methods provide a promising way. By reviewing the previous works using ML techniques for the modeling and monitoring, we observed that ML outperformed many traditional approaches for its scalability, efficiency and robustness. In future, more research with ML will be carried out for building an efficient, reliable and autonomous optical network. At the same time, there are also some challenges for ML-based techniques for practical deployments.

1. Efficient adaptation scheme. For most of the works mentioned above, the ML-based methods are trained offline with data from simulations or lab experiments before deployment. Since the weights and parameters of the ML-based methods are fixed after training, the calculation time will be short when using these methods in a practical system. This firstly-trained-then-deployed scheme is efficient for adopting ML-based methods for situations that require a fast response time. However, the data from real scenes may be different from the simulation data. Therefore, a reasonable adaptation scheme is also needed after deployment. In EON, online learning approaches such as retraining are preferable to cope with time-evolving network scenarios [73].

Even though collecting data from the practical system for retraining has been proposed in many works, the rationality for the retraining scheme needs to be reconsidered. Since the change of the EON may be unpredictable, data collected from the real scenes may not follow the same distribution with the original training data. In this case, the collected data cannot be mixed with the pre-training data to adapt the ML-based modeling/monitoring agents. Besides, if retraining agents only use the data collected from the practical system, there are other problems. On the one hand, if retraining is performed frequently for a better adaptation, dataset collected in a short period is relatively small and overfitting may occur. On the other hand, if the retraining is not frequent, estimators may have large deviations when the network state changes at a fast pace. Therefore, how to deploy an efficient adaptation scheme should be carefully considered.

2. Reasonable design of ML structure. To reach a higher accuracy, ML algorithms with more complex structures are introduced, such as DGCNN, reinforcement learning and generative adversarial network (GAN). However, these ML methods with complex structures may be hard to deploy in an optical system since they require large memories. Therefore, cost-effective ML methods are desired for EON and the structures of ML methods need to be adjusted to be tailored for the optical system.
3. Interpretability of ML-based approaches. Many works discussed in this paper are based on a neural network, which is a flexible structure for classification and regression. However, those ML algorithms often cannot provide concrete explanations for their decisions to a satisfactory extent [74]. Therefore, it is difficult to guarantee the algorithmic fairness of ML methods, which is an obstacle for deploying ML techniques to real systems. More works are desired to make ML methods interpretable to scientifically make sure that these methods can perform as expected.
4. Deployment of the ML engine. Many approaches for modeling and monitoring with ML have been proposed recently. Where to deploy these ML engines is another problem. Some ML engines can be embedded in receivers to build a low latency system while some need to be deployed in the control plane to obtain information from the whole optical networks [75]. Therefore, the strategies for the deployment of the ML engine can be carefully designed to reach an optimum performance of the ML-based method.

6. Conclusions

To improve the capacity of optical networks, planning tools with higher accuracy are required. To improve the reliability of optical networks, accurate optical performance monitoring is also desired. In this paper, we review many previous works on machine learning (ML) aided modeling and monitoring techniques in elastic optical networks. We firstly analyzed the requirements of QoT and impairment modeling. Then, by reviewing many ML-based modeling techniques, we analyze the advantages of applying ML methods for this task. Afterwards, we review and discuss various works for ML-based monitoring techniques for QoT/impairment estimation and failure management. Finally, we summarized the opportunities and challenges for the application of ML methods. Looking forward to the future, we can foresee a vital role played by ML-based mechanisms to build an intelligent optical network with high efficiency.

Author Contributions: Conceptualization, X.L., H.L., L.Y., W.H. and Q.Z.; investigation, M.F. and Y.F.; writing—original draft preparation, X.L., H.L. and Q.Z.; writing—review and editing, X.L., H.L., Y.F. and Q.Z. All authors have read and agreed to the published version of the manuscript.

Funding: This research was funded by NSFC, grant number 61801291, Shanghai Rising-Star Program, grant number 19QA1404600 and National Key R&D Program of China, grant number 2018YFB1801203.

Conflicts of Interest: The authors declare there is no conflicts of interest regarding the publication of this paper.

References

1. Berthold, J.; Saleh, A.A.M.; Blair, L.; Simmons, J.M. Optical networking: Past, present, and future. *J. Lightwave Technol.* **2008**, *26*, 1104–1118. [CrossRef]
2. Dong, Z.; Khan, F.N.; Sui, Q.; Zhong, K.; Lu, C.; Lau, A.P.T. Optical performance monitoring: A review of current and future technologies. *J. Lightwave Technol.* **2016**, *34*, 525–543. [CrossRef]
3. Pointurier, Y. Design of low-margin optical networks. *J. Opt. Commun. Netw.* **2017**, *9*, A9. [CrossRef]
4. Rottondi, C.; Barletta, L.; Giusti, A.; Tornatore, M. Machine-learning method for quality of transmission prediction of unestablished lightpaths. *J. Opt. Commun. Netw.* **2018**, *10*, A286. [CrossRef]
5. Zhuge, Q.; Zeng, X.; Lun, H.; Cai, M.; Liu, X.; Yi, L.; Hu, W. Application of machine learning in fiber nonlinearity modeling and monitoring for elastic optical networks. *J. Lightwave Technol.* **2019**, *37*, 3055–3063. [CrossRef]
6. Sinkin, O.V.; Holzlohner, R.; Zweck, J.; Menyuk, C.R. Optimization of the split-step fourier method in modeling optical-fiber communications systems. *J. Lightwave Technol.* **2003**, *21*, 61–68. [CrossRef]
7. Poggiolini, P. The GN model of non-linear propagation in uncompensated coherent optical systems. *J. Lightwave Technol.* **2012**, *30*, 3857–3879. [CrossRef]
8. Cheeseman, P.; Gevarter, W. Introduction to artificial intelligence. In Proceedings of the 24th Aerospace Sciences Meeting, American Institute of Aeronautics and Astronautics, Reno, NV, USA, 6–8 January 1986.
9. Zhang, B.; Zhang, R.; Zhang, Q.; Xin, X. Optical filtering penalty estimation using artificial neural network in elastic optical networks with cascaded reconfigurable optical add–drop multiplexers. *Opt. Eng.* **2019**, *58*, 076105.
10. Sartzetakis, I.; Christodoulopoulos, K.; Varvarigos, E. (Manos) Accurate quality of transmission estimation with machine learning. *J. Opt. Commun. Netw.* **2019**, *11*, 140. [CrossRef]
11. Svozil, D.; Kvasnicka, V.; Pospichal, J. Introduction to multi-layer feed-forward neural networks. *Chemom. Intell. Lab. Syst.* **1997**, *39*, 43–62. [CrossRef]
12. Lun, H.; Zhuge, Q.; Fu, M.; Wu, Y.; Liu, Q.; Cai, M.; Zeng, X.; Hu, W. Soft failure identification in optical networks based on convolutional neural network. In Proceedings of the 2019 European Conference on Optical Communication, Dublin, Ireland, 22–26 September 2019; p. P104.
13. Tanimura, T.; Hoshida, T.; Kato, T.; Watanabe, S.; Morikawa, H. Data-analytics-based optical performance monitoring technique for optical transport networks. In Proceedings of the Optical Fiber Communication Conference, San Diego, CA, USA, 11–15 March 2018; p. Tu3E.3.
14. Khan, F.N.; Shen, T.S.R.; Zhou, Y.; Lau, A.P.T.; Lu, C. Optical performance monitoring using artificial neural networks trained with empirical moments of asynchronously sampled signal amplitudes. *IEEE Photonics Technol. Lett.* **2012**, *24*, 982–984. [CrossRef]
15. Shariati, B.; Ruiz, M.; Comellas, J.; Velasco, L. Learning from the optical spectrum: Failure detection and identification. *J. Lightwave Technol.* **2019**, *37*, 433–440. [CrossRef]
16. Musumeci, F.; Rottondi, C.; Nag, A.; Macaluso, I.; Zibar, D.; Ruffini, M.; Tornatore, M. An overview on application of machine learning techniques in optical networks. *IEEE Commun. Surv. Tutor.* **2019**, *21*, 1383–1408. [CrossRef]
17. Khan, F.N.; Fan, Q.; Lu, C.; Lau, A.P.T. An optical communication's perspective on machine learning and Its applications. *J. Lightwave Technol.* **2019**, *37*, 493–516. [CrossRef]
18. Rafique, D.; Velasco, L. Machine learning for network automation: Overview, architecture, and applications [Invited Tutorial]. *J. Opt. Commun. Netw.* **2018**, *10*, D126. [CrossRef]
19. Mata, J.; de Miguel, I.; Durán, R.J.; Merayo, N.; Singh, S.K.; Jukan, A.; Chamania, M. Artificial intelligence (AI) methods in optical networks: A comprehensive survey. *Opt. Switch. Netw.* **2018**, *28*, 43–57. [CrossRef]
20. Fang, W.; Lu, M.; Liu, X.; Gong, L.; Zhu, Z. Joint defragmentation of optical spectrum and IT resources in elastic optical datacenter interconnections. *J. Opt. Commun. Netw.* **2015**, *7*, 314–324. [CrossRef]
21. Delezoide, C.; Christodoulopoulos, K.; Kretsis, A.; Argyris, N.; Kanakis, G.; Sgambelluri, A.; Sambo, N.; Giardina, P.; Bernini, G.; Roccato, D.; et al. Marginless operation of optical networks. *J. Lightwave Technol.* **2019**, *37*, 1698–1705. [CrossRef]
22. Channegowda, M.; Nejabati, R.; Simeonidou, D. Software-defined optical networks technology and infrastructure: Enabling software-defined optical network operations [Invited]. *J. Opt. Commun. Netw.* **2013**, *5*, A274–A282. [CrossRef]

23. Panayiotou, T.; Chatzis, S.P.; Ellinas, G. Performance analysis of a data-driven quality-of-transmission decision approach on a dynamic multicast-capable metro optical network. *J. Opt. Commun. Netw.* **2017**, *9*, 98–108. [CrossRef]
24. Politi, C.; Anagnostopoulos, V.; Matrakidis, C.; Stavdas, A. Physical layer impairment aware routing algorithms based on analytically calculated Q-factor. In Proceedings of the 2006 Optical Fiber Communication Conference and the National Fiber Optic Engineers Conference, Anaheim, CA, USA, 5–10 March 2006; p. 3.
25. Ellinas, G.; Antoniades, N.; Panayiotou, T.; Hadjiantonis, A.; Levine, A.M. Multicast routing algorithms based on Q-factor physical-layer constraints in metro networks. *IEEE Photonics Technol. Lett.* **2009**, *21*, 365–367. [CrossRef]
26. Velasco, L.; Shariati, B.; Boitier, F.; Layec, P.; Ruiz, M. Learning life cycle to speed up autonomic optical transmission and networking adoption. *J. Opt. Commun. Netw.* **2019**, *11*, 226. [CrossRef]
27. Zhang, F.; Zhuge, Q.; Plant, D.V. Fast analytical evaluation of fiber nonlinear noise variance in mesh optical networks. *J. Opt. Commun. Netw.* **2017**, *9*, C88–C97. [CrossRef]
28. Pastorelli, R.; Bosco, G.; Piciaccia, S.; Forghieri, F. Network planning strategies for next-generation flexible optical networks [Invited]. *J. Opt. Commun. Netw.* **2015**, *7*, A511–A525. [CrossRef]
29. Shao, J.; Liang, X.; Kumar, S. Comparison of split-step Fourier schemes for simulating fiber optic communication systems. *IEEE Photonics J.* **2014**, *6*, 7200515.
30. Muslu, G.M.; Erbay, H.A. Higher-order split-step Fourier schemes for the generalized nonlinear Schrödinger equation. *Math. Comput. Simul.* **2005**, *67*, 581–595. [CrossRef]
31. Musetti, S.; Serena, P.; Bononi, A. On the accuracy of split-step Fourier simulations for wideband nonlinear optical communications. *J. Lightwave Technol.* **2018**, *36*, 5669–5677. [CrossRef]
32. Murty, M.N.; Devi, V.S. *Introduction to Pattern Recognition and Machine Learning*; IISc Lecture Notes Series; Co-Published with Indian Institute of Science (IISc): Bangalore, India, 2015; Volume 5, ISBN 978-981-4335-45-4.
33. Mo, W.; Huang, Y.-K.; Zhang, S.; Ip, E.; Kilper, D.C.; Aono, Y.; Tajima, T. ANN-based transfer learning for QoT prediction in real-time mixed line-rate systems. In Proceedings of the Optical Fiber Communication Conference, San Diego, CA, USA, 11–15 March 2018; p. W4F.3.
34. Zhu, S.; Gutterman, C.L.; Mo, W.; Li, Y.; Zussman, G.; Kilper, D.C. Machine learning based prediction of Erbium-Doped fiber WDM Line amplifier gain spectra. In Proceedings of the 2018 European Conference on Optical Communication (ECOC), Rome, Italy, 23–27 September 2018; pp. 1–3.
35. You, Y.; Jiang, Z.; Janz, C. Machine learning-based EDFA gain model. In Proceedings of the 2018 European Conference on Optical Communication (ECOC), Rome, Italy, 23–27 September 2018; pp. 1–3.
36. Mo, W.; Zhu, S.; Li, Y.; Kilper, D.C. EDFA wavelength dependent gain spectrum measurement using weak optical probe sampling. *IEEE Photonics Technol. Lett.* **2018**, *30*, 177–180. [CrossRef]
37. Yan, S.; Khan, F.N.; Mavromatis, A.; Gkounis, D.; Fan, Q.; Ntavou, F.; Nikolovgenis, K.; Meng, F.; Salas, E.H.; Guo, C.; et al. Field trial of machine-learning assisted and SDN-based optical network planning with network-scale monitoring database. In Proceedings of the 2017 European Conference on Optical Communication (ECOC), Gothenburg, Sweden, 17–21 September 2017; pp. 1–3.
38. Morais, R.M.; Pedro, J. Machine learning models for estimating quality of transmission in DWDM networks. *J. Opt. Commun. Netw.* **2018**, *10*, D84. [CrossRef]
39. Wang, J.; Luo, S. Exploiting ensemble method in semi-supervised learning. In Proceedings of the 2006 International Conference on Machine Learning and Cybernetics, Dalian, China, 13–16 August 2006; pp. 1104–1107.
40. Wilcox, R.R. A note on the Theil-Sen regression estimator when the regressor Is random and the error term is heteroscedastic. *Biom. J.* **1998**, *8*, 261–268. [CrossRef]
41. Wong, S.C.; Gatt, A.; Stamatescu, V.; McDonnell, M.D. Understanding data augmentation for classification: When to warp? *arXiv* **2016**, arXiv:1609.08764.
42. Van Dyk, D.A.; Meng, X.-L. The art of data augmentation. *J. Comput. Gr. Stat.* **2001**, *10*, 1–50. [CrossRef]
43. Michie, D.; Spiegelhalter, D.J.; Taylor, C.C. *Machine Learning, Neural and Statistical Classification*; Prentice Hall: Englewood Cliffs, NJ, USA, 1994.
44. Côté, D. Using machine learning in communication networks [Invited]. *J. Opt. Commun. Netw.* **2018**, *10*, D100. [CrossRef]
45. Jia, Y.; Shelhamer, E.; Donahue, J.; Karayev, S.; Long, J.; Girshick, R.; Guadarrama, S.; Darrell, T. Caffe: Convolutional architecture for fast feature embedding. *arXiv* **2014**, arXiv:1408.5093.

46. Panayiotou, T.; Savva, G.; Shariati, B.; Tomkos, I.; Ellinas, G. Machine learning for QoT estimation of unseen optical network states. In Proceedings of the Optical Fiber Communication Conference (OFC) 2019, San Diego, CA, USA, 3–7 March 2019; p. Tu2E.2.
47. De Miguel, I.; Durán, R.J.; Jiménez, T.; Fernández, N.; Aguado, J.C.; Lorenzo, R.M.; Caballero, A.; Tafur Monroy, I.; Ye, Y.; Tymecki, A.; et al. Cognitive dynamic optical networks [Invited]. *J. Opt. Commun. Netw.* **2013**, *5*, A107. [CrossRef]
48. Jimenez, T.; Aguado, J.C.; de Miguel, I.; Duran, R.J.; Angelou, M.; Merayo, N.; Fernandez, P.; Lorenzo, R.M.; Tomkos, I.; Abril, E.J. A cognitive quality of transmission estimator for core optical networks. *J. Lightwave Technol.* **2013**, *31*, 942–951. [CrossRef]
49. Sambo, N.; Pointurier, Y.; Cugini, F.; Valcarenghi, L.; Castoldi, P.; Tomkos, I. Lightpath establishment assisted by offline QoT estimation in transparent optical networks. *J. Opt. Commun. Netw.* **2010**, *2*, 928–937. [CrossRef]
50. Meng, F.; Yan, S.; Nikolovgenis, K.; Ou, Y.; Wang, R.; Bi, Y.; Hugues-Salas, E.; Nejabati, R.; Simeonidou, D. Field trial of gaussian process learning of function-agnostic channel performance under uncertainty. In Proceedings of the Optical Fiber Communication Conference, San Diego, CA, USA, 11–15 March 2018; p. W4F.5.
51. Seve, E.; Pesic, J.; Delezoide, C.; Bigo, S.; Pointurier, Y. Learning process for reducing uncertainties on network parameters and design margins. *J. Opt. Commun. Netw.* **2018**, *10*, A298. [CrossRef]
52. Xia, L.; Zhang, J.; Hu, S.; Zhu, M.; Song, Y.; Qiu, K. Transfer learning assisted deep neural network for OSNR estimation. *Opt. Express* **2019**, *27*, 19398. [CrossRef]
53. Poggiolini, P.; Bosco, G.; Carena, A.; Curri, V.; Jiang, Y.; Forghieri, F. A simple and effective closed-form GN model correction formula accounting for signal non-gaussian distribution. *J. Lightwave Technol.* **2015**, *33*, 459–473. [CrossRef]
54. Bolshtyansky, M. Spectral hole burning in erbium-doped fiber amplifiers. *J. Lightwave Technol.* **2003**, *21*, 1032–1038. [CrossRef]
55. Woodward, S.L.; Nelson, L.E.; Feuer, M.D.; Zhou, X.; Magill, P.D.; Foo, S.; Hanson, D.; Sun, H.; Moyer, M.; O'Sullivan, M. Characterization of real-time PMD and chromatic dispersion monitoring in a High-PMD 46-Gb/s transmission system. *IEEE Photonics Technol. Lett.* **2008**, *20*, 2048–2050. [CrossRef]
56. Geyer, J.C.; Fludger, C.R.S.; Duthel, T.; Schulien, C.; Schmauss, B. Optical performance monitoring using a 43Gb/s realtime coherent receiver. In Proceedings of the 2009 IEEE/LEOS Summer Topical Meeting, Newport Beach, CA, USA, 20–22 July 2009; pp. 93–94.
57. Tanimura, T.; Hoshida, T.; Rasmussen, J.C.; Suzuki, M.; Morikawa, H. OSNR monitoring by deep neural networks trained with asynchronously sampled data. In Proceedings of the 2016 21st OptoElectronics and Communications Conference (OECC) Held Jointly with 2016 International Conference on Photonics in Switching (PS), Niigata, Japan, 3–7 July 2016.
58. Wang, D.; Zhang, M.; Li, J.; Li, Z.; Li, J.; Song, C.; Chen, X. Intelligent constellation diagram analyzer using convolutional neural network-based deep learning. *Opt. Express* **2017**, *25*, 17150–17166. [CrossRef]
59. Tanimura, T.; Hoshida, T.; Kato, T.; Watanabe, S.; Morikawa, H. Convolutional neural network-based optical performance monitoring for optical transport networks. *J. Opt. Commun. Netw.* **2019**, *11*, A52. [CrossRef]
60. Tan, M.C.; Khan, F.N.; Al-Arashi, W.H.; Zhou, Y.; Tao Lau, A.P. Simultaneous optical performance monitoring and modulation format/bit-rate identification using principal component analysis. *J. Opt. Commun. Netw.* **2014**, *6*, 441–448. [CrossRef]
61. Khan, F.N.; Yu, Y.; Tan, M.C.; Yu, C.; Lau, A.P.T.; Lu, C. Simultaneous OSNR monitoring and modulation format identification using asynchronous single channel sampling. In Proceedings of the Asia Communications and Photonics Conference 2015, Hong Kong, China, 19–23 November 2015; p. AS4F.6.
62. Khan, F.N.; Zhong, K.; Zhou, X.; Al-Arashi, W.H.; Yu, C.; Lu, C.; Lau, A.P.T. Joint OSNR monitoring and modulation format identification in digital coherent receivers using deep neural networks. *Opt. Express* **2017**, *25*, 17767. [CrossRef] [PubMed]
63. Anderson, T.B.; Kowalczyk, A.; Clarke, K.; Dods, S.D.; Hewitt, D.; Li, J.C. Multi Impairment monitoring for optical networks. *J. Lightwave Technol.* **2009**, *27*, 3729–3736. [CrossRef]
64. Wang, C.; Fu, S.; Xiao, Z.; Tang, M.; Liu, D. Long short-term memory neural network (LSTM-NN) enabled accurate optical signal-to-noise ratio (OSNR) monitoring. *J. Lightwave Technol.* **2019**, *37*, 4140–4146. [CrossRef]
65. Wang, Z.; Yang, A.; Guo, P.; He, P. OSNR and nonlinear noise power estimation for optical fiber communication systems using LSTM based deep learning technique. *Opt. Express* **2018**, *26*, 21346. [CrossRef]

66. Skoog, R.A.; Banwell, T.C.; Gannett, J.W.; Habiby, S.F.; Pang, M.; Rauch, M.E.; Toliver, P. Automatic identification of impairments using support vector machine pattern classification on eye diagrams. *IEEE Photonics Technol. Lett.* **2006**, *18*, 2398–2400. [CrossRef]
67. Kashi, A.S.; Zhuge, Q.; Cartledge, J.; Borowiec, A.; Charlton, D.; Laperle, C.; O'Sullivan, M. Artificial neural networks for fiber nonlinear noise estimation. In Proceedings of the Asia Communications and Photonics Conference, Guangzhou, China, 10–13 November 2017; p. Su1B.6.
68. Vaquero Caballero, F.J.; Ives, D.; Zhuge, Q.; O'Sullivan, M.; Savory, S.J. Joint estimation of linear and non-linear signal-to-noise ratio based on neural networks. In Proceedings of the Optical Fiber Communication Conference, San Diego, CA, USA, 11–15 March 2018; p. M2F.4.
69. Zhang, K.; Fan, Y.; Ye, T.; Tao, Z.; Oda, S.; Tanimura, T.; Akiyama, Y.; Hoshida, T. Fiber nonlinear noise-to-signal ratio estimation by machine learning. In Proceedings of the Optical Fiber Communication Conference (OFC) 2019, San Diego, CA, USA, 3–7 March 2019; p. Th2A.45.
70. Vela, A.P.; Ruiz, M.; Fresi, F.; Sambo, N.; Cugini, F.; Meloni, G.; Poti, L.; Velasco, L.; Castoldi, P. BER degradation detection and failure identification in elastic optical networks. *J. Lightwave Technol.* **2017**, *35*, 4595–4604. [CrossRef]
71. Shahkarami, S.; Musumeci, F.; Cugini, F.; Tornatore, M. Machine-learning-based soft-failure detection and identification in optical networks. In Proceedings of the Optical Fiber Communication Conference, San Diego, CA, USA, 11–15 March 2018; p. M3A.5.
72. Varughese, S.; Lippiatt, D.; Richter, T.; Tibuleac, S.; Ralph, S.E. Identification of soft failures in optical links using low complexity anomaly detection. In Proceedings of the Optical Fiber Communication Conference (OFC) 2019, San Diego, CA, USA, 3–7 March 2019; p. W2A.46.
73. Musumeci, F.; Rottondi, C.; Corani, G.; Shahkarami, S.; Cugini, F.; Tornatore, M. A tutorial on machine learning for failure management in optical networks. *J. Lightwave Technol.* **2019**, *37*, 4125–4139. [CrossRef]
74. Vellido, A.; Martın-Guerrero, J.D.; Lisboa, P.J.G. Making machine learning models interpretable. *Comput. Intell.* **2012**, *12*, 163–172.
75. Boitier, F.; Layec, P. Automated optical networks with monitoring and machine learning. In Proceedings of the 2018 20th International Conference on Transparent Optical Networks (ICTON), Bucharest, Romania, 1–5 July 2018; pp. 1–4.

applied sciences

MDPI

Article

A Simple Joint Modulation Format Identification and OSNR Monitoring Scheme for IMDD OOFDM Transceivers Using K-Nearest Neighbor Algorithm

Qianwu Zhang, Hai Zhou, Yuntong Jiang, Bingyao Cao, Yingchun Li, Yingxiong Song, Jian Chen *, Junjie Zhang and Min Wang

Key Laboratory of Specialty Fiber Optics and Optical Access Networks, Shanghai Institute for Advanced Communication and Data Science, Shanghai University, Shanghai 200072, China; zhangqianwu@shu.edu.cn (Q.Z.); zhouhai3053@shu.edu.cn (H.Z.); jiangtoyun@shu.edu.cn (Y.J.); cby85064@163.com (B.C.); liyingchun@shu.edu.cn (Y.L.); herosf@shu.edu.cn (Y.S.); zjj@staff.shu.edu.cn (J.Z.); wangmin@mail.shu.edu.cn (M.W.)

* Correspondence: chenjian@staff.shu.edu.cn

Received: 26 June 2019; Accepted: 10 September 2019; Published: 17 September 2019

Featured Application: This paper provides a feasible method for modulation format identification and OSNR monitoring. The application of KNN reduces complexity.

Abstract: In this study, a joint modulation format identification and optical signal-to-noise ratio (OSNR) monitoring algorithm is proposed and experimentally demonstrated using the k-nearest neighbor algorithm for intensity modulation and direct detection (IMDD) orthogonal frequency division multiplexing (OFDM) systems. A modified amplitude histogram of received signal is employed to serve as the classification feature to simplify the computation. Experimental results show that five common quadrature amplitude modulation (QAM) modulation formats, including 4-QAM, 16-QAM, 32-QAM, 64-QAM and 128-QAM, can be identified under 100% accurate estimation at the received optical power of −11 dBm. Robustness of the proposed scheme to constellation rotation is also experimentally assessed. At the same time, system OSNR monitoring also can be achieved and the average prediction mean square error (MSE) is 0.69 dB^2, which is similar to that using an artificial neural network. Computational complexity assessment demonstrated that similar performance but less computing resource consumption can be achieved by using the proposed scheme rather than the artificial neural network-based scheme.

Keywords: k-nearest neighbor algorithm; modulation format identification; OSNR monitoring

1. Introduction

The elastic optical networks [1] (EONs), as one of the solutions of the next-generation fiber-optic communication network, have recently attracted a great deal of attention. Optical orthogonal frequency division multiplexing (OOFDM) is considered a promising alternative to the EONs because of its advantages, including its high spectral efficiency, good resistance to chromatic dispersion, as well as its provision of hybrid dynamic bandwidth allocation in both frequency and time domains. IMDD realizations, e.g., directly modulated laser (DML) based OOFDM transceivers [2], are usually applied in cost-sensitive systems, e.g., optical access networks or data centers, because of their simple structure and DSP (digital signal processing) [3]. With adaptively modulated optical orthogonal frequency division multiplexing (AMOOFDM) [4–9], systems can maximize capacity and achieve flexible bandwidth allocation simultaneously. However, to achieve the bandwidth on demand (BoD) approach, the EONs are expected to be able to dynamically adjust various transmission parameters, e.g., modulation formats, OSNR, spectrum assignments, etc., depending on the adjusting traffic demands and network condition. Negotiations between the transmitters and receivers are vital for adjusting the modulation format or other parameters before resetting the bit rate. To achieve the flexible bandwidth provision, one of the critical functional requirements for intelligent receivers in EONs is the ability to identify the modulation formats and monitor OSNR of received signals without any prior information through the negotiations between transmitters and receivers.

Modulation format identification has aroused growing attention from researchers in recent years. Bo investigated and characterized a blind modulation format recognition method by projecting partial received data in Stokes space onto a 2-D plane to plot a converted binary graph [10]. Based on the evaluation of the peak-to-average-power ratio of the incoming data samples, Bilal presented a simple novel digital modulation format identification scheme for coherent optical systems after some independent DSP processing at the receiver [11]. Liu proposed a modulation format identification technique based on the extraction and identification of specific features of received signal power distributions in digital coherent receivers, which successfully identified six modulation formats [12]. However, the method described in [10] requires the approximate OSNR and the carrier frequency of the received signals, the method in [11] requires additional hardware components as well as the OSNR of received signals, and the scheme in [12] requires the computation of the received signal power distributions for digital coherent receivers.

In recent years, several machine learning-based modulation format identification (MFI) techniques have been proposed both in digital coherent and directly detected receivers [13–22] for optical communications systems because of their excellent learning ability from data, which can avoid the requirement of pre-information. Khan proposed a deep machine learning method to identify three modulation formats at an accuracy of 100% in a wide optical signal-to-noise ratio range [13]. A simple and cost-effective MFI technique was also proposed by his teams using an artificial neural network based on asynchronous amplitude histogram (AAH) [14]. Guesmi experimentally demonstrated a cost-effective technique to achieve optical performance monitoring functionalities and enable simultaneous symbol rate and modulation format identification based on artificial neural networks [15]. Jiang introduced a novel modulation format identification method based on intensity fluctuation features using support vector machines [16]. Zhang utilized an artificial neural network to identify modulation format and a genetic algorithm to simplify the structures of an artificial neural network for directly detected receivers [17]. However, these methods focus merely on identifying the modulation format of the received signal and do not provide any information about the quality of signal in terms of OSNR. Joint OSNR monitoring and modulation format identification in digital coherent receivers using deep neural network (DNN) was performed in [18]. The excellent experimental effect of DNN was considered, but the problem of high complexity was neglected.

In this study, a joint modulation format identification and OSNR monitoring algorithm is proposed and experimentally demonstrated using the k-nearest neighbor (KNN) algorithm for IMDD OFDM systems. A modified amplitude histogram (AH) of received signal is employed to serve as the classification feature to simplify the computation. Experimental results show that five common QAM modulation formats including 4-QAM, 16-QAM, 32-QAM, 64-QAM and 128-QAM can be identified under 100% accurate estimation at the received optical power of −11 dBm. Robustness of the proposed scheme to constellation rotation is also experimentally assessed. At the same time, system OSNR monitoring can also be achieved with an average prediction MSE of 0.69 dB^2, which is similar to what is achieved using an artificial neural network. Computational complexity assessment demonstrated that similar performance but less computing resource consumption can be achieved by using the proposed scheme rather than the artificial neural network (ANN)-based scheme. AMOOFDM without transceiver negotiations can also be achieved using the proposed scheme, showing its good potential for intelligent transceivers in elastic optical networks.

2. Operation Principle of Proposed KNN Based Scheme

The KNN algorithm [23] was first proposed by Cover and Hart, which is commonly employed for text categorization, a process of identifying the class to which a text document belongs. The KNN algorithm is considered a simple and intuitive algorithm: comparing the features of a testing data set with a training data set and finding the k instances closest to the instance in the training dataset under a given training dataset, for the new input testing instances. Moreover, the category corresponding to testing data is the one with the largest number of occurrences in k instances. The operation procedure of the KNN algorithm is described as follows:

Firstly, the input dataset is written as:

$$T = \{(x_1, y_1), (x_2, y_2), \ldots, (x_N, y_N)\}, i = 1, 2, \ldots, N \tag{1}$$

where $x_i \in \chi \subseteq R^n$ denotes the feature vector of the instance; $y_i \in Y = \{c_1, c_2, \ldots, c_k\}$ is the category of the instance. The distance L between input training data and testing data is calculated as Equation (2):

$$\mathrm{L}(x_i, x_j) = \left(\sum_{l=1}^{2} \left|x_i^{\,l} - x_j^{\,l}\right|^2\right)^{\frac{1}{2}} = \sqrt{\left|x_i^{(1)} - x_j^{(1)}\right|^2 + \left|x_i^{(2)} - x_j^{(2)}\right|^2} \tag{2}$$

where i and j denote the instance of training data and testing data, l denotes the dimension.

Subsequently, the distance is sorted according to the rising relation of distance.

Next, following the given distance measurement, the k nearest points to x are found in the training set T before determining the occurrence frequency of the categories of the preceding k points. And the neighborhoods of x covering the k points are expressed as $N_k(x)$, which is used in Equation (3).

Finally, the category with the highest frequency in the preceding k points is returned as the prediction classification of testing data. The majority voting strategy is selected as the decision criteria:

$$\mathrm{y} = \underset{c_j}{\arg max} \sum_{x_i \in N_k(x)} I(y_i = c_j), i = 1, 2, \ldots, N; j = 1, 2, \ldots, K \tag{3}$$

where I denotes the indicator function. When $y_i = c_j$, I is 1, otherwise 0.

As shown in Figure 1, the green dot represents the test data, blue squares represent category A of the training set, and red triangles represent category B of the training set. If the value of k is set to 3, the number of red triangles will be 2 greater than the blue squares, of which there is 1. Thus, the category of green dot is red triangle. However, when the value of k increases to 5, the category of green dot belongs to blue square since the number of blue squares is greater than the red triangles. The special case of the k-nearest neighbor algorithm is k = 1, which is named as the nearest neighbor algorithm.

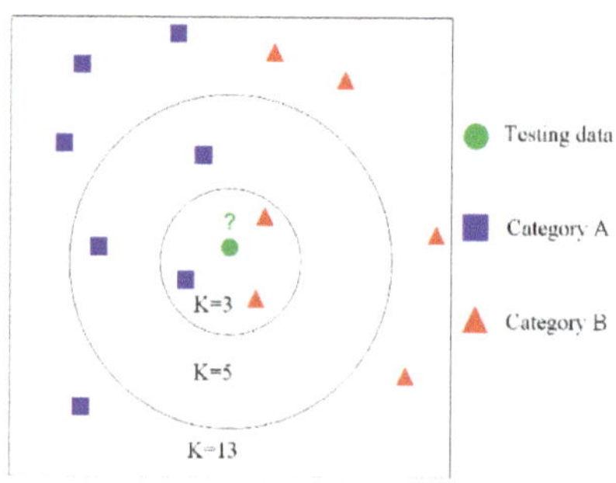

Figure 1. The k-nearest neighbor (KNN) schematic diagram.

There are two situations dealing with the complex data. One is the training stage in which modulation format of training data is known, and the other is the testing stage, which is the unknown data waiting for identification with the KNN classifier obtained in the training stage. For the training data, the KNN algorithm is utilized to process the feature vector to build a model which is virtually the KNN classifier. For the unknown data, the feature vector is fed into the KNN classifier, and the modulation format of this data can be obtained. It is worth mentioning that KNN does not have an explicit learning process. In fact, it is a well-known representative of lazy learning. This kind of learning technology only saves samples at the training stage. After receiving the test sample, the subsequent processing is carried out, including comparing the test data with the stored data.

In the receivers of the noted OOFDM systems, fast Fourier transform (FFT) processing is first performed on the received signal to obtain the complex signal data, and then the features of the data can be abstracted from the complex signal data. The AH is introduced in the proposed scheme in which only the real part of the received signals is involved to abstract features. In this study, 30 samples are selected as a training data set, of which every sample is a subcarrier transmission data of OFDM signals. Subsequently, for every subcarrier, 3000 data points are selected from whole 20,400 data points to obtain features. Accordingly, for the 4-QAM, 16-QAM, 32-QAM, 64-QAM and 128-QAM, 2, 4, 6, 8 and 12 peaks should be observed in corresponding AHs. The clustering degree of peaks represents different OSNR. Obviously, when the peak is prominent, the value of OSNR is large.

To simplify the computation complexity and improve the robustness of the algorithm, a data preprocessing scheme is first implemented. The real part of the complex data is divided into 100 intervals from −1.5 to 1.5, after taking real part operations for each data point. The number of points falling in different intervals is used as input feature vector data for KNN. To reduce the effect that computational overhead increases with the increase in data size, each interval is processed as follow:

(a) If the number of points in an interval ≥256, the value of this interval will be set to 32.
(b) If the number of points in an interval ≤32, the value of this interval will be set to 0.
(c) If the number of points in an interval is between 32 and 256, the value of this interval will be set to an integer obtained by using the number divided by 8, which can be conveniently performed by a commercial processor in the future.

Then, the final features of each interval can be obtained.

3. Experimental Verification and Discussions

An IMDD OFDM transmission system over an SSMF (standard single mode fiber) as illustrated in Figure 2 is employed to experimentally evaluate the performance of the proposed scheme [24]. Detailed transceiver and system key parameters can be found in Table 1. In this system, all DSP procedures for both transmitter and receiver are achieved by an offline approach. At the transmitter side, the input pseudo random data (PRBS15) is first mapped into parallel complex data with five modulation formats from 4-QAM to 128-QAM. A 64-point inverse fast Fourier transform (IFFT) module is then applied for the generation of OFDM time-domain symbols, in which 30 of them can be used to allocate user data to satisfy the Hermitian symmetry for a real-valued IMDD signal approach. Next, an arbitrary waveform generator (AWG) is employed to generate analog signals at a sampling rate of 2 GSa/s. The analog OFDM signals are then fed into the preamplifiers to adjust the signal amplitude before directly driving a DFB laser at 1550 nm with bandwidth of 2 GHz. Subsequently, the DFB laser converts the electrical signals to optical signals and sends them to the 25 km SSMF.

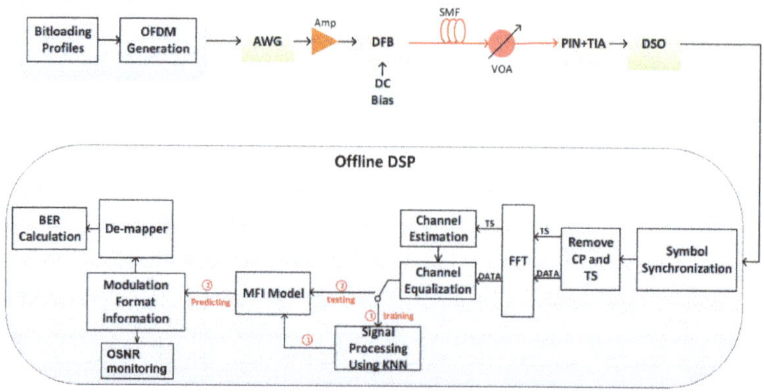

Figure 2. The experimental setup of intensity modulation and direct detection (IMDD) orthogonal frequency division multiplexing (OFDM) system.

Table 1. Transceiver and system parameters.

Parameter	Value
Modulation format	128/64/32/16/4-QAM
IFFT/FFT size	64 points
CP length	16 points
Bit rate	4.28 Gb/s
OFDM symbols per frame	100 symbols
SSMF length	25 km
DFB modulation bandwidth	2 GHz
DFB wavelength	1550 nm
PIN detector bandwidth	12 GHz
AWG sampling rate	2 GS/s
DSO sampling rate	10 GS/s
DSO/AWG resolution	8 bit
Training set sample size	30 sample
Data points per sample	3000 points

At the received side, a 12 GHz PIN (positive intrinsic-negative) with trans-impedance amplifier (TIA) is utilized for O-E conversion by directly detecting the optical OFDM signals. The received optical power (RoP) can be adjusted by a variable optical attenuator (VOA). The received signals are then captured by the digital storage oscilloscope (DSO) with 10 GSa/s sampling rate ADC to convert analog signals to digital signals for offline DSP processing.

During the offline DSP processing, the received data is converted to complex data after symbol synchronization, cyclic prefix (CP)/training sequence (TS) removal, FFT, and channel equalization, as shown in Figure 2. The model of KNN is obtained by training, and then the modulation format is predicted by this model. Once the modulation format is realized using the proposed scheme, the complex data can be decoded to get the original transmit data and to calculate the BER (bit error rate). Meanwhile, OSNR is predicted based on the corresponding modulation formats. The specific recognition and prediction algorithm based on KNN is explained in detail in Section 2.

To investigate the feasibility of the proposed scheme, subcarrier bit loading profiles corresponding to a total bit rate of 4.28 Gb/s shown in Figure 3 are employed for the measurements, including five different mixed QAM modulation formats from 4-QAM to 128-QAM. In this paper, we intentionally chose the bit loading profiles to ensure that every type of modulation format was introduced in the experiment. In fact, in practical application, the proposed algorithm will certainly be used with an adaptive bit-and-power-loading algorithm. The constellations and corresponding abstracted representative AHs when the RoP is −11 dBm are shown in Figure 4a. The Figure suggests that different peak profiles can be observed for different modulation formats. We can see from Figure 4a that the constellation maps of 4-QAM and 16-QAM have good performance, and that the modulation formats can be easily distinguished. When it comes to 32-QAM, the constellation becomes blurred, but obvious peak value can be seen by the AH method, which means that the modulation format can be well recognized, as shown in Figure 4b. When the modulation formats are 64-QAM and 128-QAM, not only the constellation map is blurred, but also there is no obvious peak value in the AH map. Therefore, KNN is used to process data to get the correct modulation format. As shown in Figure 4b, the shape of AH when RoP is −11 dBm is different from that when RoP is −6 dBm. Thus, as long as the corresponding OSNR under each RoP is known, OSNR value can be predicted according to different AH under different RoP. Consider that KNN has the function of regression prediction [25,26], which can be used in achieving OSNR monitoring [18].

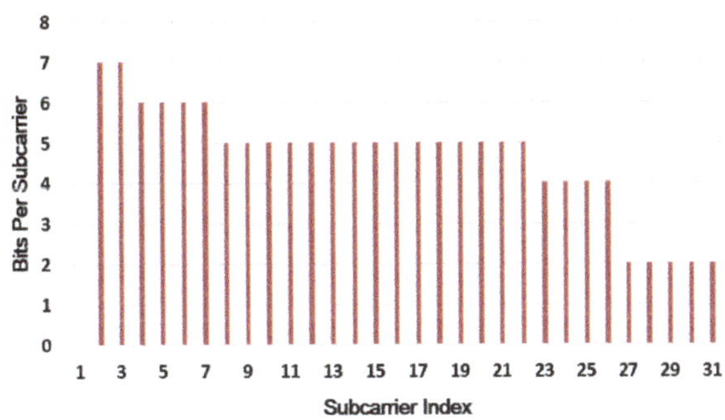

Figure 3. Subcarrier bit allocation profile for orthogonal frequency division multiplexing (OFDM) signals.

Since identification accuracy is affected by the number of training samples and the *k* value of the KNN algorithm, two different sets of samples and *k* value, being 30, 1 (case 1) and 60, 3 (case 2) are introduced in the construction of the classifier. The accuracy identification for 4 different sets of bit loading profiles while maintaining the same bit rate under case 1 and case 2 are shown in Figure 5. In general, the accuracy increases with the increase in the number of samples and the *k* value. Besides, there is no exception in this measurement. However, the improvement in accuracy is not significantly obvious as the performance of case 2 (dotted line) is only slightly improved at the RoP of <−12 dBm and cannot reach 100%. Once the RoP reaches −12 dBm or higher, the performance of the two mentioned cases will be the same. This suggests that the number of training samples and the *k* value are not the main influencing factors of the proposed algorithm. To increase the transmission efficiency, the number of training samples is 30, and the *k* value is 1 in the proposed KNN training model during the measurements. As shown in Figure 4, when RoP is −11 dBm, the constellation map is already blurred. However, the identification accuracy of this algorithm can reach 100%, which shows that the algorithm also has a good recognition effect in the case of a relatively high error rate.

Subsequently, the above measurement is repeated under different RoP for both optical back to back (OBTB) and 25 km SSMF configurations. The experimental results involving the mentioned 4 sets of different bit loading profiles are shown in Figure 6, in which the BERs under each bit loading profile is also plotted. Note that the identification accuracy of proposed MFI increases with the increase in RoP. Almost the same performance is obtained for 25 km SSMF transmissions compared to the BTB case, which shows that fiber dispersion does not significantly affect BER and the proposed MFI for current system configuration. This is mainly because our focus in this paper is on the algorithm verification for different types of modulation format, the signal rate involved in our experiment is not very high, and the chromatic dispersion induced system degradation is relatively small. For both 25 km SSMF and OBTB cases, 100% accurate identification can be achieved when the RoP is higher than −11 dBm.

For the RoP >−10 dBm, the BER performance of OBTB and 25 km SSMF can be lower than the adopted HD-FEC limit of 3.8×10^{-3} and power penalty is <0.5 dB for 4 different bit loading profiles. System performance with longer distance and higher signal rate will be studied in future works.

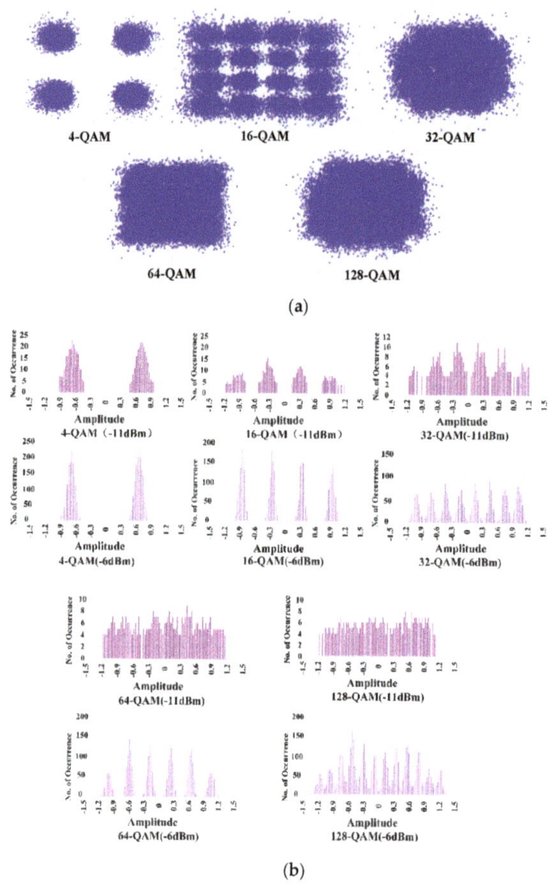

Figure 4. (**a**) The constellations, (**b**) AH (RoP = −11 dBm and −6 dBm) of mixed QAM modulation format from 4-QAM to 128-QAM.

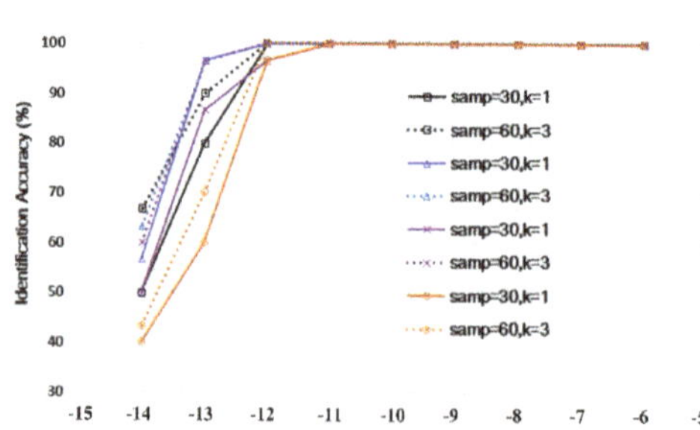

Figure 5. The comparison of identification accuracy with different samples and k value.

From Figure 6, identification accuracy depends on the receiving optical power, that is, the signal-to-noise ratio. Employment of a stronger equalization algorithm can improve the BER performance and also should improve the identification accuracy at low optical power regions. In practical deployment, the appropriate algorithm can be selected according to the requirements to achieve a balance between performance and complexity.

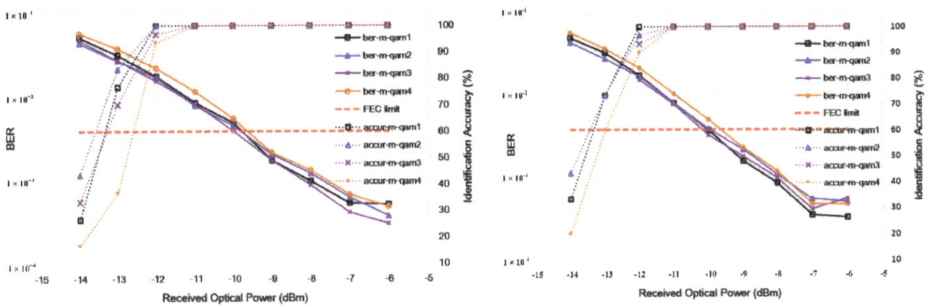

(**a**) (**b**)

Figure 6. The BER curves and identification accuracy for (**a**) OBTB and (**b**) 25 km SSMF configurations.

In practical application, constellation rotation after equalization affects the performance of MFI performance. Further experimental investigation is undertaken to assess the effects of constellation rotation. The bit loading as shown in Figure 3 is employed, and additional constellation rotation of $\pi/64$, $\pi/48$, $\pi/32$ and $\pi/28$ is added during the measurements by adjusting the initial phase of the QAM constellation in offline DSP procedures of transmitter. The identification accuracies under different additional phase rotations are shown in Figure 7. It is noteworthy that the identification accuracy increases with the increase in RoP. For the additional phase rotation of $\pi/32$, 100% accuracy cannot be achieved until the RoP meets −6 dBm for the 25 km SSMF case and −7 dBm for the OBTB case. However, for the additional phase rotation of $\pi/28$, 100% accuracy cannot be achieved even under RoP of −6 dBm. For the additional phase rotation of $<\pi/32$, identification accuracy is nearly unchanged compared with the configuration without additional phase rotation, suggesting good robustness to residual constellation rotation.

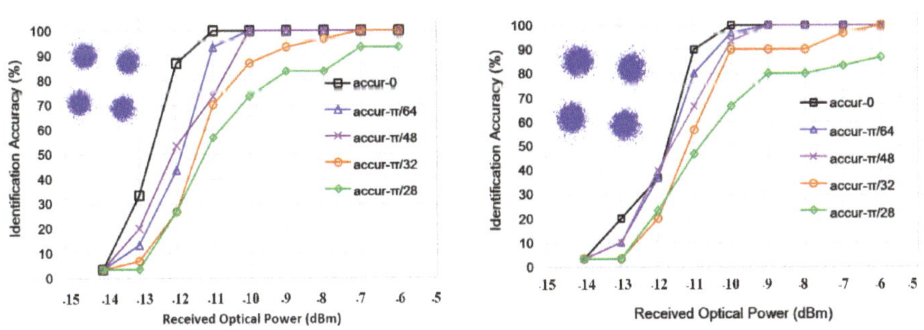

(**a**) (**b**)

Figure 7. The identification accuracy under different residual phase rotation for (**a**) OBTB and (**b**) 25 km SSMF configurations. 4-QAM rotated constellation diagram for the additional phase rotation of $\pi/32$ at a specific RoP (−11 dBm) is embedded.

When it comes to OSNR monitoring, the k value should also be determined first. As shown in Figure 8a, as the k value increases, the difference between the estimated and true OSNR increases in general. Specifically, when the k value is small, the estimated OSNR is approximately proportional to the true OSNR. However, when the k value is large, the estimated OSNR is almost unaffected by the true. The MSE between estimated and true OSNR is then calculated to determine the value of k. The result is as shown in Figure 8b. When $k = 2$, MSE is the smallest, so k is set to 2 when performing OSNR monitoring.

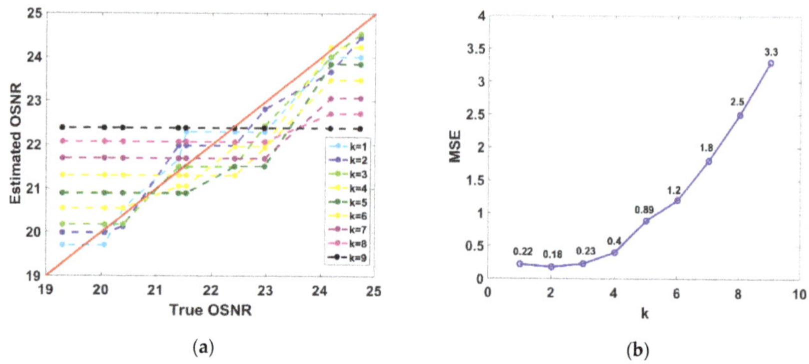

Figure 8. (**a**) Estimated versus true OSNRs with different k values and (**b**) MSE versus k values for 128 QAM (OBTB).

The OSNR monitoring results for five signal types are shown in Figure 9. It is clear from the figure that OSNR estimates are quite accurate for both OBTB and 25 km SSMF configurations. Hence, the mean OSNR estimation error for the three signal types considered in this work is 0.69 dB, which is similar to the ones reported for the OSNR monitoring [18]. It is worth mentioning that the proposed algorithm can also work for coherent optical systems and single-carrier modulation.

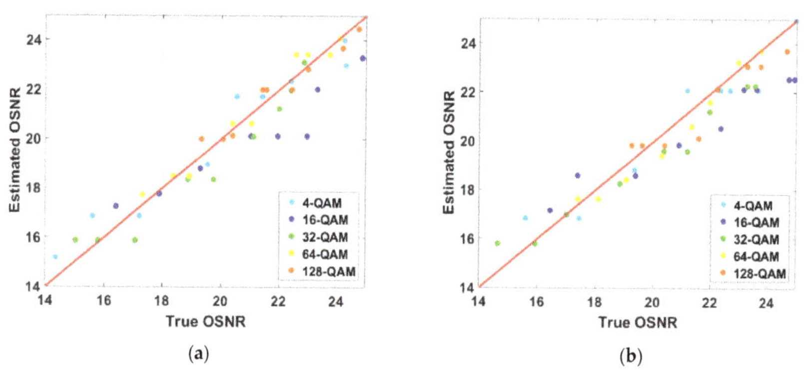

Figure 9. Estimated versus true OSNRs for (**a**) OBTB and (**b**) 25 km SSMF configurations.

We also compared the complexity of the proposed KNN algorithm with artificial neural network [13]. The complexity calculation of ANN [27,28] and KNN are listed in Table 2, in which the complexity calculation involves two parts, namely the training part and the prediction part. For ANN, N_{ep} is the number of samples in a training set and n_i, n_{hid} and n_o are the number of neurons on the input, hidden and output layers, respectively. For KNN, N_{TS} is the number of samples in a training set and n is the number of features. The identification accuracy results of both algorithms under the same condition are shown in Figure 10 and the identification accuracy is similar. MSE of OSNR monitoring for both algorithms when RoP is −11 dBm is listed in Table 3. The average MSE of KNN algorithm is 0.69 dB^2, and that of the ANN algorithm is 0.71 dB^2. In general, compared to the ANN, the KNN algorithm can effectively reduce multiplication operations. However, the KNN algorithm has similar performance to the ANN algorithms.

Table 2. The complexity comparison between KNN and ANN.

Algorithm	Multiplications	MFI	OSNR Monitoring
ANN	$C_{train} = N_{ep}(n_i n_{hid} + n_{hid} n_o)$ $C_{predict} = n_i n_{hid} + n_{hid} n_o$ $C_{ANN} = C_{train} + C_{predict}$	$C_{train} = 100 \times (101 \times 4 + 40 \times 5) = 424,000$ $C_{test} = 101 \times 40 + 40 \times 5 = 4240$ $C_{ANN} = 428,240$	$C_{train} = 100 \times (101 \times 40 + 40) = 408,000$ $C_{test} = 101 \times 40 + 40 = 4080$ $C_{ANN} = 412,080$
KNN	$C_{KNN} = C_{train} + C_{predict} = 2C_{train} = 2N_{TS}n$	$C_{KNN} = 2 \times 30 \times 101 = 6060$	$C_{KNN} = 2 \times 30 \times 101 = 6060$

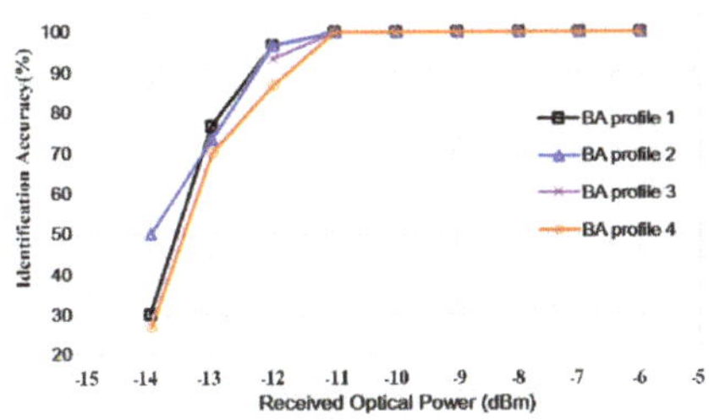

(**a**)

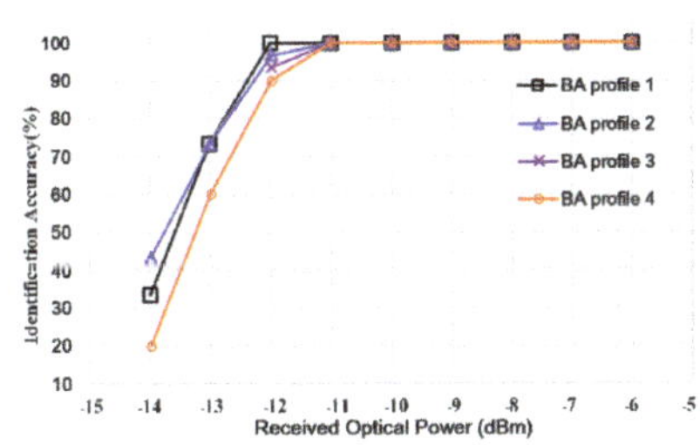

(**b**)

Figure 10. The identification accuracy of (**a**) ANN algorithm and (**b**) KNN algorithm.

Table 3. The MSE of KNN and ANN algorithm.

MSE	4-QAM	16-QAM	32-QAM	64-QAM	128-QAM
KNN(dB^2)	0.66	1.3	0.77	0.3	0.42
ANN(dB^2)	0.87	0.65	0.69	0.84	0.49

4. Conclusions

In this study, a joint modulation format identification and OSNR—monitoring algorithm is proposed and experimentally demonstrated using the KNN algorithm for IMDD OFDM systems. A modified AH of received signal is employed to serve as the classification feature to simplify the computation. According to the experimental results, five common QAM modulation formats can be identified with a 100% accurate estimation at the RoP of −11 dBm. At the same time, system OSNR monitoring also can be achieved with an average prediction MSE of 0.69 dB^2, which is similar to that using artificial neural network. Robustness and computational complexity of the proposed scheme are also experimentally assessed.

Author Contributions: Q.Z. put forward the research and innovation points of the article and revised the paper. H.Z. carried out experimental verification and prepared the paper. Y.J., B.C., Y.L., Y.S., J.C., J.Z. and M.W. help provide funding and experimental environments.

Funding: This research was funded by National Natural Science Foundation of China (Project No. 61601279, 61420106011, 61601277, 61635006); Shanghai Science and Technology Development Funds (Project No. 17010500400, 16511104100).

Conflicts of Interest: There is no conflict of interest.

References

1. Gerstel, O.; Jinn, M.; Lord, A.; Yoo, S.J.B. Elastic optical networking: A new dawn for the optical layer? *IEEE Commun. Mag.* **2012**, *50*, S12–S20. [CrossRef]
2. Tang, J.M.; Lane, P.M.; Shore, K.A. High-Speed Transmission of Adaptively Modulated Optical OFDM Signals Over Multimode Fibers Using Directly Modulated DFBs. *J. Lightw. Technol.* **2006**, *24*, 429. [CrossRef]
3. Tang, J.M.; Giddings, R.P.; Jin, X.Q.; Wei, J.L.; Zheng, X.; Giacoumidis, E.; Hugues-Salas, E.; Hong, Y.; Shu, C.; Groenewald, J.; et al. Real-Time Optical OFDM Transceivers for PON Applications. In Proceedings of the 2011 Optical Fiber Communication Conference and Exposition and the National Fiber Optic Engineers Conference, Los Angeles, CA, USA, 6–10 March 2011. Paper OTuK3.
4. Zheng, X.; Tang, J.M.; Spencer, P.S. Transmission Performance of Adaptively Modulated Optical OFDM Modems Using Subcarrier Modulation over Worst-Case Multimode Fibre Links. *IEEE Commun. Lett.* **2008**, *12*, 788–790. [CrossRef]
5. Zheng, X.; Wei, J.; Tang, J.M. Transmission performance of adaptively modulated optical OFDM modems using subcarrier modulation over SMF IMDD links for access and metropolitan area networks. *Opt. Express* **2008**, *16*, 20427–20440. [CrossRef] [PubMed]
6. Duong, T.N.; Genay, N.; Ouzzif, M.; Masson, J.L.; Charbonnier, B.; Chanclou, P.; Simon, J.C. Adaptive Loading Algorithm Implemented in AMOOFDM for NG-PON System Integrating Cost-Effective and Low-Bandwidth Optical Devices. *IEEE Photonic Technol. Lett.* **2009**, *21*, 790–792. [CrossRef]
7. Giacoumidis, E.; Wei, J.L.; Jin, X.Q.; Tang, J.M. Improved transmission performance of adaptively modulated optical OFDM signals over directly modulated DFB laser-based IMDD links using adaptive cyclic prefix. *Opt. Express* **2008**, *16*, 9480–9494. [CrossRef] [PubMed]
8. Zhao, J.; Ellis, A.D. A novel optical fast OFDM with reduced channel spacing equal to half of the symbol rate per carrier. In Proceedings of the Optical Fiber Communication Conference 2010, San Diego, CA, USA, 21–25 March 2010.
9. Xing, O.; Jian, Z. Single Side-Band Modulated Coherent Optical Fast-OFDM with Frequency Oversampling. *J. Lightw. Technol.* **2017**, *35*, 2436–2443.
10. Bo, T.; Tang, J.; Chan, C.K. Modulation Format Recognition for Optical Signals Using Connected Component Analysis. *IEEE Photonic Tech. Lett.* **2017**, *29*, 11–14. [CrossRef]
11. Bilal, S.M.; Bosco, G.; Dong, Z.H.; Lau, A.P.T.; Lu, C. Blind modulation format identification for digital coherent receivers. *Opt. Express* **2015**, *23*, 26769–26778. [CrossRef] [PubMed]
12. Liu, J.; Dong, Z.H.; Zhong, K.P.; Lau, A.P.T.; Lu, C.; Lu, Y.Z. Modulation Format Identification Based on Received Signal Power Distributions for Digital Coherent Receivers. In Proceedings of the European Conference on Optical Communication, Cannes, France, 21–25 September 2014. Paper Th4D.3.
13. Khan, F.N.; Zhong, K.; Al-Arashi, W.H.; Yu, C.; Lu, C.; Lau, A.P.T. Modulation Format Identification in Coherent Receivers Using Deep Machine Learning. *IEEE Photonic Technol. Lett.* **2016**, *28*, 1886–1889. [CrossRef]
14. Khan, F.N.; Zhou, Y.; Lau, A.P.T.; Lu, C. Modulation format identification in heterogeneous fiber-optic networks using artificial neural networks. *Opt. Express* **2012**, *20*, 12422–12431. [CrossRef] [PubMed]
15. Guesmi, L.; Ragheb, A.M.; Fathallah, H.; Menif, M. Experimental Demonstration of Simultaneous Modulation Format/Symbol Rate Identification and Optical Performance Monitoring for Coherent Optical Systems. *J. Lightw. Technol.* **2018**, *36*, 2230–2239. [CrossRef]
16. Jiang, L.; Yan, L.; Yi, A.; Pan, Y.; Bo, T.; Hao, M.; Pan, W.; Luo, B. Robust and Blind Modulation Format Identification for Elastic Optical Networks. In Proceedings of the Optical Fiber Communication Conference, San Francisco, CA, USA, 9–13 March 2014; pp. 1–3.

17. Zhang, S.; Peng, Y.; Shu, Q.; Li, J.; Li, Z. Modulation format identification in heterogeneous fiber-optic networks using artificial neural networks and genetic algorithms. *Photonic Netw. Commun.* **2016**, *32*, 246–252. [CrossRef]
18. Khan, F.N.; Zhong, K.; Zhou, X.; Al-Arashi, W.H.; Yu, C.; Lu, C.; Lau, A.P.T. Joint OSNR monitoring and modulation format identification in digital coherent receivers using deep neural networks. *Opt. Express* **2017**, *25*, 17767. [CrossRef] [PubMed]
19. Tan, M.C.; Khan, F.N.; Al-Arashi, W.H.; Zhou, Y.; Lau, A.P.T. Simultaneous optical performance monitoring and modulation format/bit-rate identification using principal component analysis. *J. Opt. Commun. Netw.* **2014**, *6*, 441–448. [CrossRef]
20. Kim, B.; Kim, J.; Chae, H.; Yoon, D.; Choi, J.W. Deep neural network-based automatic modulation classification technique. In Proceedings of the International Conference on Information and Communication Technology Convergence, Jeju Island, Korea, 19–21 October 2016; IEEE: Piscataway, NJ, USA, 2016; pp. 579–582.
21. Khan, F.N.; Yu, Y.; Tan, M.C.; Al-Arashi, W.H.; Yu, C.; Lau, A.P.T.; Lu, C. Experimental demonstration of joint OSNR monitoring and modulation format identification using asynchronous single channel sampling. *Opt. Express* **2015**, *23*, 30337–30346. [CrossRef] [PubMed]
22. Cheng, L.; Xi, L.; Zhao, D.; Tang, X.; Zhang, W.; Zhang, X. Improved modulation format identification based on Stokes parameters using combination of fuzzy c-means and hierarchical clustering in coherent optical communication system. *Chin. Opt. Lett.* **2015**, *13*, 100604. [CrossRef]
23. Cover, T.; Hart, P. Nearest neighbor pattern classification. *IEEE Trans. Inf. Theory* **1967**, *13*, 21–27. [CrossRef]
24. Zhang, Q.; Liu, M.; Chen, J.; Dun, H.; Zhang, J.; Li, Y.; Song, Y.; Wang, M. Real-time optical OFDM transmissions with spectral efficiency up to 6.93 bit/s/Hz over 50 km SSMF IMDD systems. *Opt. Commun.* **2017**, *387*, 12–17. [CrossRef]
25. Lian, H. Convergence of functional k-nearest neighbor regression estimate with functional responses. *Electron. J. Stat.* **2011**, *5*, 31–40. [CrossRef]
26. Farahnakian, F.; Pahikkala, T.; Liljeberg, P.; Plosila, J. Energy Aware Consolidation Algorithm Based on K-Nearest Neighbor Regression for Cloud Data Centers. In Proceedings of the IEEE/ACM International Conference on Utility & Cloud Computing, Washington, DC, USA, 9–12 December 2013.
27. Kyono, T.; Otsuka, Y.; Fukumoto, Y.; Owaki, S.; Nakamura, M. Computational-Complexity Comparison of Artificial Neural Network and Volterra Series Transfer Function for Optical Nonlinearity Compensation with Time- and Frequency-Domain Dispersion Equalization. In Proceedings of the European Conference on Optical Communication, Roma, Italy, 23–27 September 2018; pp. 1–3.
28. Sidelnikov, O.; Redyuk, A.; Sygletos, S. Nonlinear Equalization in Long Haul Transmission Systems Using Dynamic Multi-Layer Perceptron Networks. In Proceedings of the European Conference on Optical Communication, Roma, Italy, 23–27 September 2018; pp. 1–3.

applied sciences

MDPI

Article

Photon Enhanced Interaction and Entanglement in Semiconductor Position-Based Qubits

Panagiotis Giounanlis [1,†], Elena Blokhina [1,†,*], Dirk Leipold [2] and Robert. Bogdan Staszewski [1,2]

1 School of Electrical and Electronic Engineering, University College Dublin, Dublin 4, Ireland; panagiotis.giounanlis@ucd.ie (P.G.); robert.staszewski@ucd.ie (R.B.S.)

2 Equal1. Labs, Fremont, CA 94536, USA; dirk.leipold@equal1.com

* Correspondence: elena.blokhina@ucd.ie

† These authors contributed equally to this work.

Received: 6 September 2019; Accepted: 18 October 2019; Published: 25 October 2019

Abstract: CMOS technologies facilitate the possibility of implementing quantum logic in silicon. In this work, we discuss a minimalistic modelling of entangled photon communication in semiconductor qubits. We demonstrate that electrostatic actuation is sufficient to construct and control desired potential energy profiles along a Si quantum dot (QD) structure allowing the formation of position-based qubits. We further discuss a basic mathematical formalism to define the position-based qubits and their evolution under the presence of external driving fields. Then, based on Jaynes–Cummings–Hubbard formalism, we expand the model to include the description of the position-based qubits involving four energy states coupled with a cavity. We proceed with showing an anti-correlation between the various quantum states. Moreover, we simulate an example of a quantum trajectory as a result of transitions between the quantum states and we plot the emitted/absorbed photos in the system with time. Lastly, we examine the system of two coupled position-based qubits via a waveguide. We demonstrate a mechanism to achieve a dynamic interchange of information between these qubits over larger distances, exploiting both an electrostatic actuation/control of qubits and their photon communication. We define the entanglement entropy between two qubits and we find that their quantum states are in principle entangled.

Keywords: entanglement; charge qubit; position-based semiconductor qubits; cryogenic technologies; semiconductor photon communication; Jaynes–Cummings–Hubbard formalism

1. Introduction

In a world where the quantum technology rapidly emerges, CMOS technologies seem to be attractive candidates for a successful widely spread realization of quantum computers. Recently, efforts have been reported towards the construction of quantum computers on a silicon chip [1–4]. When cooled down to cryogenic temperatures, CMOS allows constructing effective quantum potential energy profiles where particles can realize their nano-world properties. Such implementations can be based on spin-spin interaction or in the position of single and/or multiple fermions or even Majorana particles [5–10].

The manipulation of semiconductor qubits can be carried out in principle by the application of an electric/magnetic field. The control over semiconductor position-based qubits (also known in the literature as charge qubits) can be achieved by using only a static electric field, which is of interest in this work. It can be shown that one can move the charge between neighboring quantum dots (QD) by applying appropriate voltages across a given array of a quantum register/structure. The fine lithography of nanometer-scale CMOS allows the use of single-electron devices (SED) for the injection and detection of single electrons, as well as their qunatum transport through the quantum register.

Correlated oscillations and signatures of entanglement of silicon-based qubits have been demonstrated in the literature, making the construction of quantum gates and algorithms possible [11–15].

At the same time, the possibility of integration of double-quantum-dots (DQDs) with microwave cavities was recently demonstrated, where an absorbed photon facilitates tunneling of electrons between QDs by means of pumping energy into the system. However, a reversed phenomenon is also possible, i.e., a photon emission due to the tunneling of particles in a quantum transport operation, where the DQDs are coupled through a cavity [16–21]. Previous experiments showed that the cavity photon emission can be enhanced as a result of entangled states between coupled atoms. It is also known that semiconductor quantum dots, acting as artificial atoms, can emit photons, opening a possibility to achieve communication between qubits within a chip. This photon exchange via a waveguide is expected to enhance the coupling between DQDs by introducing a superposition of many quantum paths [22–25]. By proper manipulation of the system and by exploiting its optical properties, one could control the transport of flying qubits and, as a result, the dynamic interchange of information.

In our previous work, we showed that it is possible to control and construct semiconductor position-based qubits in DQDs in nanometer-scale CMOS. We also showed that electrostatically coupled interacting DQDs are in principle entangled [26,27]. In this paper, motivated by the recent demonstrations of semiconductor-based photon communication, and expanding our previous work on modelling of position-based semiconductor qubits, we will focus on a minimalistic modelling of two-coupled DQDs in silicon interconnected with a waveguide.

To start, we illustrate that electrostatic actuation is sufficient for the construction of valid potential energy profiles. This allows the formation of position-based qubits in a DQD. We show how one can apply voltages and voltage pulses to manipulate the tunneling rates and energy levels in a quantum register.

We continue by describing a system of a position-based qubit exhibiting four energy levels, coupled with a cavity. We demonstrate photon emission and absorption through quantum paths by capturing the dynamics of the various quantum states of the system. Remarkably, anticorrelations of quantum states can be observed, even in a single-qubit system.

Finally, we expand our description to a system of two coupled position-based qubits via a waveguide. We define the Von Neumann entanglement entropy of the system, and then show that anticorrelation and entanglement between the quantum states of the two qubits can be achieved through an interchange of photons. The electrostatic actuation of the position-based qubits coupled through a cavity allows quantum communication via longer distances, opening the possibility of the formation of quantum gates between non-neighboring qubits.

The paper is organized as follows: Section 2 discusses the system under study. Section 3 introduces a brief description of the position-based qubit and the basic transitions between the quantum states that can take place in the system. Section 4 develops a minimalistic modelling based on the Jaynes–Cummings–Hubbard formalism. We examine selected case studies and present simulation results demonstrating photon absorption, emission and entanglement.

2. Statement of the Problem

The quantum structure under study is realized in a CMOS fully depleted silicon-on-insulator (FDSOI) technology. A transistor-like CMOS device is depicted in Figure 1a. When operating at cryogenic temperatures, such devices can be seen as artificial atoms, giving rise to a double-quantum-dot (DQD). DQDs can be arranged in series forming a quantum register of an arbitrary number of qubits. The quantum transport can be achieved electrostatically by manipulating the electrical potential, i.e., by applying appropriate DC and pulsed voltages at the gates. For a more detailed description of this system, an interested reader can refer to [26].

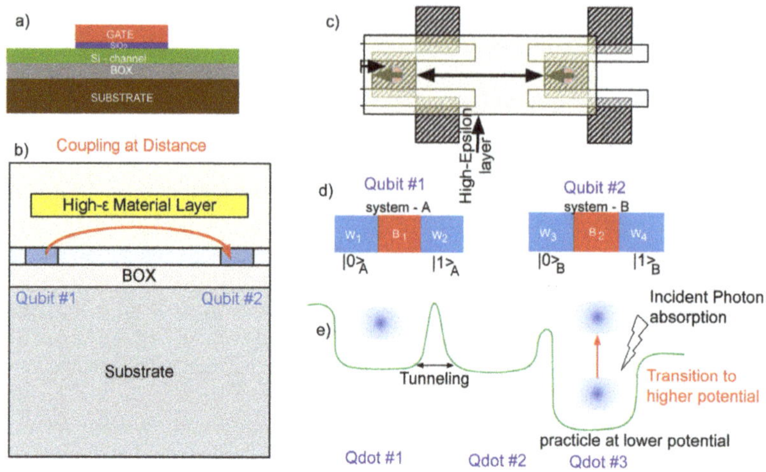

Figure 1. (**a**) Cross-section of an FDSOI transistor-like 'quantum' device, without the source/drain diffused regions. Details of the specific technology are omitted. As the device dimensions decrease, one can achieve quantum operation at cryogenic temperatures. Various layers of the device facilitate different properties which are essential to the resulting potential energy profile. (**b**) Each device can facilitate a double-quantum-dot (DQD). When interconnected via a waveguide, it can couple the entangled qubits. This can be achieved with the use of high-ϵ materials and proper isolation of the quantum core from the rest of the surrounding circuitry of a chip. (**c**) Top-view of a layout structure of two coupled DQDs interconnected via a waveguide. (**d**) 1D representation of two coupled (DQDs). Each DQD can facilitate a qubit. In the schematic, the two states of each qubit are denoted, as $|0\rangle_{A,B}$, corresponding to a particle in the left QD of system-A or system-B (of the corresponding potential energy profile of a device), and as $|1\rangle_{A,B}$, corresponding to a particle in the right QD. (**e**) 1D schematic representation of a potential energy profile formed by a chain of devices forming a series of QDs. The absorption of a photon can cause a transition to a higher energy level. Similarly, the emission of a photon can occur when a particle transitions from a higher energy level to a lower energy level.

In Figure 1b, the concept of coupling two qubits over longer distances, in the case electrostatic (Coulomb) interaction is negligible, is demonstrated. The coupling is expected to be achieved through a waveguide by means of a high-ϵ material layer. Figure 1c visualizes a top-view layout of two coupled DQDs with a waveguide. Each DQD realizes a qubit. In this work, we will focus on the modelling of a qubit coupled into a cavity and two coupled qubits coupled via a waveguide. As visualized in Figure 1d, each qubit has two states, denoted as $|0\rangle$ and $|1\rangle$. These are localized states which correspond to the presence of a particle in a given well of an electrostatically shaped potential energy profile. In other words, the quantum logic in such a system is based on the position of the particle in a given quantum register.

By manipulating the potential energy of a particle one can achieve photon emission and absorption. An example of the potential energy on the surface of silicon channel of a quantum register, as obtained from electrostatic finite-element-method (COMSOL) simulations, is shown in Figure 2a. At the beginning (and also at the end) of the chain, the first (and the last) device serves as a reconfigurable injector/detector. The injector, in this context, is typically a single-electron device responsible for injecting individual electrons into the quantum register. The detector is an analog circuit which can detect (by a weak or strong measurement, depending on a particular configuration) a presence of a single electron. In the schematic, "S" and "D" denote Source and Drain regions (i.e., highly doped silicon), while "I_1–I_7" denote imposers (gates). One can apply precisely controlled DC voltages and pulses to the source/drain and the imposers to achieve desired modes of operation.

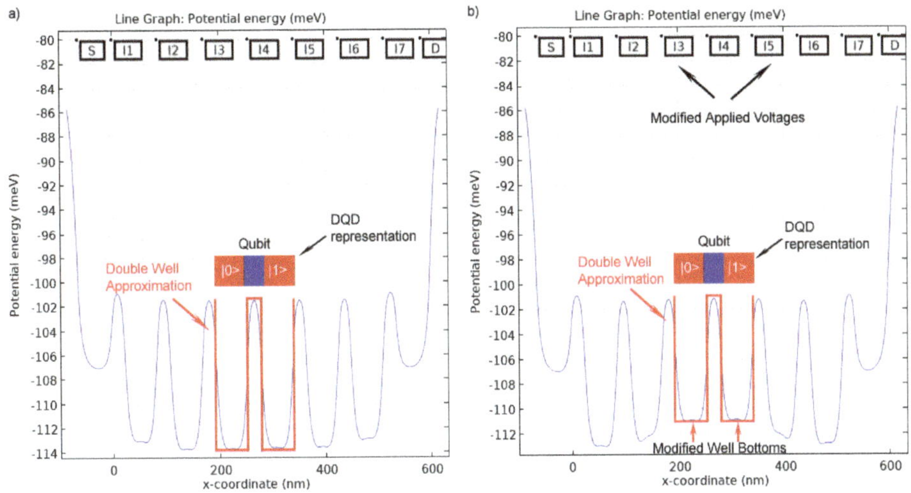

Figure 2. (**a**) Finite element method (FEM) COMSOL simulations of the electrostatically shaped potential energy as a function of distance assuming carrier freezout operation for a chain of six devices for a quantum register. Appropriate voltage configuration allows one to construct a desired potential energy profile of a desired mode of operation. In the figures, "S/D" denote the source/drain, while "I_1–I_7"denote the imposers (gates). It can be seen that the potential energy profile can be approximated by an equivalent square-potential energy profile. The position-based qubit can be defined in a region of two potential wells separated by a barrier. For example, as shown, a qubit can be defined between the imposers "I_3–I_5". The double well can be approximated by a DQD, The smaller the dimensions of the structure the more accurate this approximation. (**b**) By manipulating V_{GS} (gate-source) and V_{DS} (drain-source) applied voltages, one can achieve various potential energy profiles. In the schematic, the potential well bottoms between the imposers "I_3–I_5" are raised, which is equivalent to a lowered energy barrier between them. With the use of such an electrostatic mechanism, one can manipulate the resulting tunneling probability between the wells. Such an alteration of a potential energy profile can be achieved in a specific implementation by keeping V_{GS}, V_{DS} voltages constant and by applying pulses at the imposers at precisely controlled time instances, duration and magnitude. Depending on the magnitude of the pulse, one can pump enough energy into the system so the particle (assuming in the ground state) gets excited from the ground energy and jumps to a new energy level. These transitions between energy levels due to the perturbed driving field can cause photon emission, typically of the same energy as the perturbation. This mechanism is similar to the absorption of a photon from a cavity. In this case, if the photon's energy is similar to the gap between the two energy levels in a potential well, the particle (assuming already in the ground state) can become excited to allow tunneling.

In such a geometry, with specifically voltages applied, barriers (or wells) are formed between (or under) the imposers. In principle, the resulting potential energy profile can be approximated by an equivalent square potential energy profile. The DQD, defined as an abstract approximation of a double-well, i.e., in this example in the region between any two imposers I_3–I_5, can represent a position-based qubit. This will be discussed in more detail in the next section.

It is also possible to slightly alter the potential energy profile with purely electrostatic actuation, i.e., by applying positive or negative voltage pulses of a specific width and magnitude at the gate(s). The outcome is schematically illustrated in Figure 2b, where the well bottoms between imposers "I_3–I_5" are raised, which is equivalent to a lowered barrier between the wells. With such an electrostatic mechanism, electron tunneling can be induced or prohibited (more accurately, the probability of tunneling can be increased or decreased). In other words, by lowering (raising) the barrier, the probability of tunneling increases (decreases) drastically. Depending on the magnitude of perturbation, one can excite a particle by pumping energy into the system. From this point of view, an electrostatic driving field/perturbation can have a similar result as photon absorption (electromagnetic driving field) from a cavity. A tunneling electron will then emit photons of specific (discrete) energies related to the bounded energy levels of the constructed potential wells.

To sum up, as with the electrostatic actuation, photon absorption can also activate tunneling. Therefore, as discussed and demonstrated later in this paper, the coupling between DQDs via a waveguide allows the manipulation of qubits, exploiting both the electrostatic manipulation and photon exchange.

3. Position-Based Semiconductor Qubits in the Frame of Semiconductor Photon Communication

3.1. Rabi Flopping Frequency of a Position-Based Qubit

We start our analysis by considering the system of a CMOS double-quantum-dot (DQD). We assume initially that only the two lowest states can be populated. In this case, the system can be seen as a two-level system: the ground state, denoted as $|g\rangle$, and the first excited state, denoted as $|e\rangle$.

In this basis, the wave-function can be expressed as a superposition of the two eigen-functions:

$$|\psi\rangle = c_g\,|\psi_g\rangle + c_e\,|\psi_e\rangle \tag{1}$$

where c_g and c_e are the probability amplitudes of each energy state, with $c_g^2 + c_e^2 = 1$. The energy difference between the ground state, E_g, and the first excited state, E_e, defines the occupancy frequency, $\omega_0 = (E_e - E_g)/\hbar$. When in relaxation, the system is expected to be in the ground state. However, one can excite the system by applying a driving field, which can be a square or a sinusoidal pulse of a given width. Assuming initially a small pertubative pulse, the system will be in a quantum superposition of the first two states, and it will display occupancy oscillations [26].

The Hamiltonian of the system can be written as:

$$\hat{H} = \hat{H}_0 + \hat{H}_I \tag{2}$$

where $\hat{H}_0$ is the Hamiltonian of the system in equilibrium, with $\hat{H}_0 = \hat{p}^2/2m_e^* + U_0$, $\hat{H}_I = U_{BI}\cos\omega t$ is the interaction Hamiltonian and m_e^* is the effective electron mass. U_{BI} denotes the change to the initial potential energy profile, U_0, of the DQD system, when an arbitrary bias voltage waveform of frequency ω is applied at the gates (The interaction Hamiltonian due to a single-mode cavity field of frequency ω will have the same effect and can be represented qualitatively by the same expression). In the rotating wave approximation, the Rabi flopping frequency, ω_R, is [28]:

$$\omega_R \equiv \frac{1}{2}\sqrt{(\omega-\omega_0)^2 + (U_{BI}/\hbar)^2} \tag{3}$$

and

$$\begin{aligned} c_e(t) &= i\tfrac{U_{BI}}{\Omega_{\mathrm{R}}\pi} e^{i\delta t/2}\sin\left(\omega_{\mathrm{R}}t/2\right) \\ c_g(t) &= e^{i\delta t/2}\left\{\cos\left(\omega_{\mathrm{R}}t/2\right) - i\tfrac{\delta}{\omega_{\mathrm{R}}}\sin\left(\omega_{\mathrm{R}}t/2\right)\right\} \end{aligned} \tag{4}$$

where $\delta = \omega - \omega_0$ is the detuning term. In this work, we are not interested in investigating the effects of detuning, therefore we assume $\delta = 0$. The transition probability, $P_{g\to e} \equiv P_e(t)$, can be written as:

$$P_e(t) = |C_e(t)|^2 \quad = \frac{U_{BI}^2}{\omega_{\mathrm{R}}^2\pi^2}\sin^2\left(\omega_{\mathrm{R}}t/2\right) \tag{5}$$

3.2. Representation of the System in a Position Basis

Qualitatively, we can write the Hamiltonian in the tight binding approximation:

$$H = \begin{vmatrix} E_p & t_{s,0\to 1} \\ t_{s,1\to 0} & E_p \end{vmatrix} \tag{6}$$

where $t_{s,i\to j} = \langle i|\hat{H}|j\rangle, i \neq j = \{0,1\}$ are the tunneling (also known as hoping) terms in the tight-binding formalism which express the tunneling probability between the two neighboring quantum dots. Now, if we set $E_g = E - \Delta$ and $E_e = E + \Delta$, then $E_e - E_g = 2\Delta$ and $\omega_0 = \frac{2\Delta}{\hbar}$, so (6) can be written as [29,30]:

$$H = \begin{vmatrix} E & -\Delta \\ -\Delta & E \end{vmatrix} \tag{7}$$

with eigen-values $\lambda = E \pm \Delta$. The wave-function, in a position basis can be written as:

$$|\phi\rangle = c_0 |0\rangle + c_1 |1\rangle \tag{8}$$

where c_0 and c_1 are the occupancy coefficients in the Wannier-position basis $\{|0\rangle,|1\rangle\}$, with

$$\begin{aligned} |0\rangle\,(t=0) &= \frac{1}{\sqrt{2}}(|\psi_g\rangle + |\psi_e\rangle) \\ |1\rangle\,(t=0) &= \frac{1}{\sqrt{2}}(|\psi_g\rangle - |\psi_e\rangle) \end{aligned} \tag{9}$$

The time evolution of the system, starting from state $|0\rangle\,(t=0)$ is:

$$\begin{aligned} |\phi\rangle\,(t) = \frac{1}{\sqrt{2}}(|\psi_g\rangle\, e^{-i(E-\Delta)t/\hbar} + |\psi_e\rangle\, e^{-i(E+\Delta)t/\hbar} = \\ \frac{1}{\sqrt{2}}\left(\frac{1}{\sqrt{2}}\Big(|0\rangle + |1\rangle\Big)e^{-i(E-\Delta)t/\hbar} + \frac{1}{\sqrt{2}}\Big(|0\rangle - |1\rangle\Big)e^{-i(E+\Delta)t/\hbar}\right) = \\ \frac{1}{\sqrt{2}}e^{-iEt/\hbar}\left(e^{i\Delta t/\hbar}\Big(|0\rangle + |1\rangle\Big) + (e^{-i\Delta t/\hbar}\Big(|0\rangle - |1\rangle\Big)\right) = e^{-iEt/\hbar}\left(\cos(\frac{\Delta t}{\hbar})\,|0\rangle + i\cdot\sin(\frac{\Delta t}{\hbar})\,|1\rangle\right) \end{aligned} \tag{10}$$

The above time-dependent oscillations have frequency probabilities:

$$\begin{aligned} P_{|0\rangle}(t) &= |\langle 0|0\rangle\,(t)\rangle|^2 = \cos^2\left(\tfrac{\Delta t}{\hbar}\right) \\ P_{|1\rangle}(t) &= |\langle 1|1\rangle\,(t)\rangle|^2 = \sin^2\left(\tfrac{\Delta t}{\hbar}\right) \end{aligned} \tag{11}$$

Notice that we can write:

$$|\phi(t)\rangle = c_0\,|0\rangle\, e^{i\omega_0 t} + c_1\,|1\rangle\, e^{-i\omega_0 t} \tag{12}$$

4. Photon Emission Due to Transitions in a Semiconductor Position-Based Qubit—Description Based on a Jaynes–Cummings–Hubbard Formalism

4.1. Description of the System of Coupled Position-Based Qubit with a Cavity

In this section, we will expand the system of semiconductor position-based qubit by including a Jaynes–Cummings–Hubbard formalism for the full description of the system of a coupled qubit with a cavity. We will additionally assume four energy levels, and a square double-well potential energy approximation for the potential energy profile of the double quantum dot, as they are depicted in Figure 3. All the simulations in this study were performed assuming the potential energy profile and energy levels of Figure 3. In this case, the four bound states of the system can be expressed in the position basis $\{|0_g\rangle|n\rangle, |1_g\rangle|n\rangle, |0_e\rangle|n+1\rangle, |1_e\rangle|n+1\rangle\}$:

$$|0_g\rangle|n\rangle = \frac{1}{\sqrt{2}}\Big(|\psi_0\rangle|n\rangle + |\psi_1\rangle|n\rangle\Big), \quad |1_g\rangle|n\rangle = \frac{1}{\sqrt{2}}\Big(|\psi_0\rangle|n\rangle - |\psi_1\rangle|n\rangle\Big)$$
$$|0_e\rangle|n+1\rangle = \frac{1}{\sqrt{2}}\Big(|\psi_2\rangle|n+1\rangle + |\psi_3\rangle|n+1\rangle\Big), \quad |1_e\rangle|n+1\rangle = \frac{1}{\sqrt{2}}\Big(|\psi_2\rangle|n+1\rangle - |\psi_3\rangle|n+1\rangle\Big) \tag{13}$$

where the localized $|0,1\rangle_{g,e}$ states are expressed as a superposition of $|\psi_i\rangle, i = 0,\ldots,3$ eigen-states, and $n \in \mathbb{N}$ denotes the number of photons. In this basis, the Hamiltonian of the system in spectral representation can be written as:

$$\hat{H} = \sum_{i=\{0,1\},k=\{g,e\}} E_{p,k}|i_k\rangle\langle i_k| + \sum_{i\neq j=\{0,1\},k=\{g,e\}} t_{s,i_k\to j_k}|i_k\rangle\langle j_k| + \sum_{i=\{0,1\},k\neq l=\{g,e\}} t_{c,i_k\to i_l}|i_k\rangle\langle i_l| \tag{14}$$

where E_p denotes the potential energy terms of each site, t_s denotes the hopping terms of the tunneling between neighbor quantum dots of the same energy level, $|g\rangle$ or $|e\rangle$, and t_c denotes the frequency of transitions between the lower $|g\rangle$ and upper $|e\rangle$ energy states of the same localized state, $|0\rangle$ or $|1\rangle$. We should also note that we consider photon emission only in transitions from the ground energy split (the first two eigen-energies, E_0, E_1) to the excited energy split (eigen-energies E_2, E_3) without losing generality of our results. Then, from (14), the time-evolution of the system can be found from the time-dependent Schrödinger equation in a matrix representation:

$$i\hbar\frac{\mathrm{d}c_m}{\mathrm{d}t} = \sum_n H_{mn}c_n \tag{15}$$

where c_m are the probability amplitudes of each coefficient of the superposition of quantum states. For the given initial conditions, a solution of (15) will have the form:

$$|\psi(t)\rangle = c_{0g}(t)|\psi_0\rangle + c_{0e}(t)|\psi_1\rangle + c_{1g}(t)|\psi_3\rangle + c_{1e}(t)|\psi_4\rangle \tag{16}$$

where in this notation, for example, amplitude c_{0g} corresponds to the ground localized state, $|0_g\rangle$, i.e., where "0" is the position (state $|0\rangle$) and "g" is the energy level (state $|g\rangle$). Then, the time evolution of the probability of the particle to be found, for example in the state $|0_g\rangle$, will be denoted as:

$$P_{|0_g\rangle}(t) = |c_{0g}(t)|^2 \tag{17}$$

or in general

$$P_{|n_m\rangle}(t) = |c_{nm}(t)|^2, \; n = 0,1 \; m = g,e \tag{18}$$

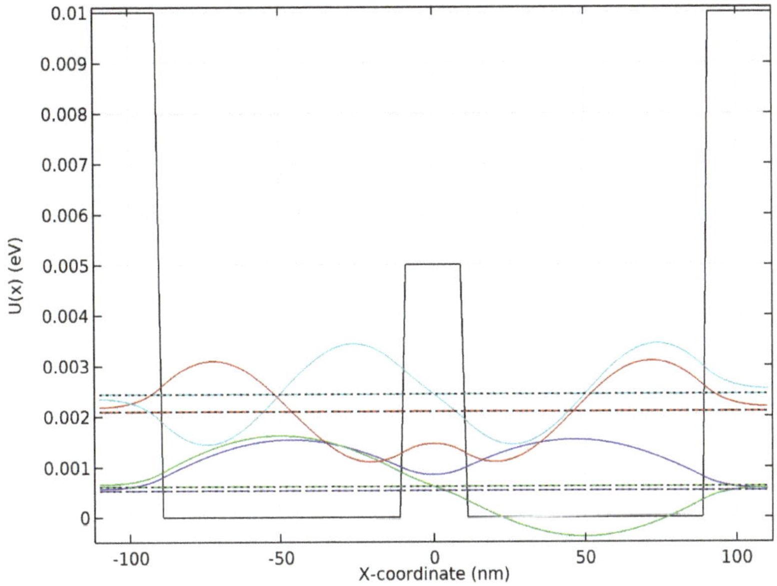

Figure 3. Potential energy profile as a piecewise approximation of a double-well with four energy-levels. This function is extracted from COMSOL electrostatic simulations, assuming carrier freezout operation. The number of bound states in the double well depends on a particular voltage configuration applied at the gates. Therefore, it is possible to electrostatically control the energy gaps between the different energy levels. In addition, the energy levels of a given potential energy double well determine the allowed energies of photons that are possible to emitte and absorb during the quantum operation between communicating qubits.

4.2. Simulation Results

The time evolution of the probabilities of different quantum states, as obtained from solving (15), is depicted in Figure 4. The probability of the states $|0_g\rangle$ and $|0_e\rangle$ is visualized in Figure 4a. The probabilities of the localized states $|0\rangle$ are, as expected, anti-correlated. Similarly, the probabilities $|1_g\rangle$ and $|1_e\rangle$ are also anti-correlated, as shown in Figure 4b. Therefore, the evolution of quantum states in terms of ground $|g\rangle$ and excited $|e\rangle$ energies exhibits an anti-correlation. Moreover, in Figure 4c one can see that the anti-correlation occurs also in the probabilities of the two localized states $|0_g\rangle$ and $|1_g\rangle$, of a quantum particle of the same ground level, $|g\rangle$, i.e., this is a positional anti-correlation. The same positional anti-correlation occurs between the two localized states, $|0_e\rangle$ and $|1_e\rangle$, but now of the same excited energy level, as depicted in Figure 4d.

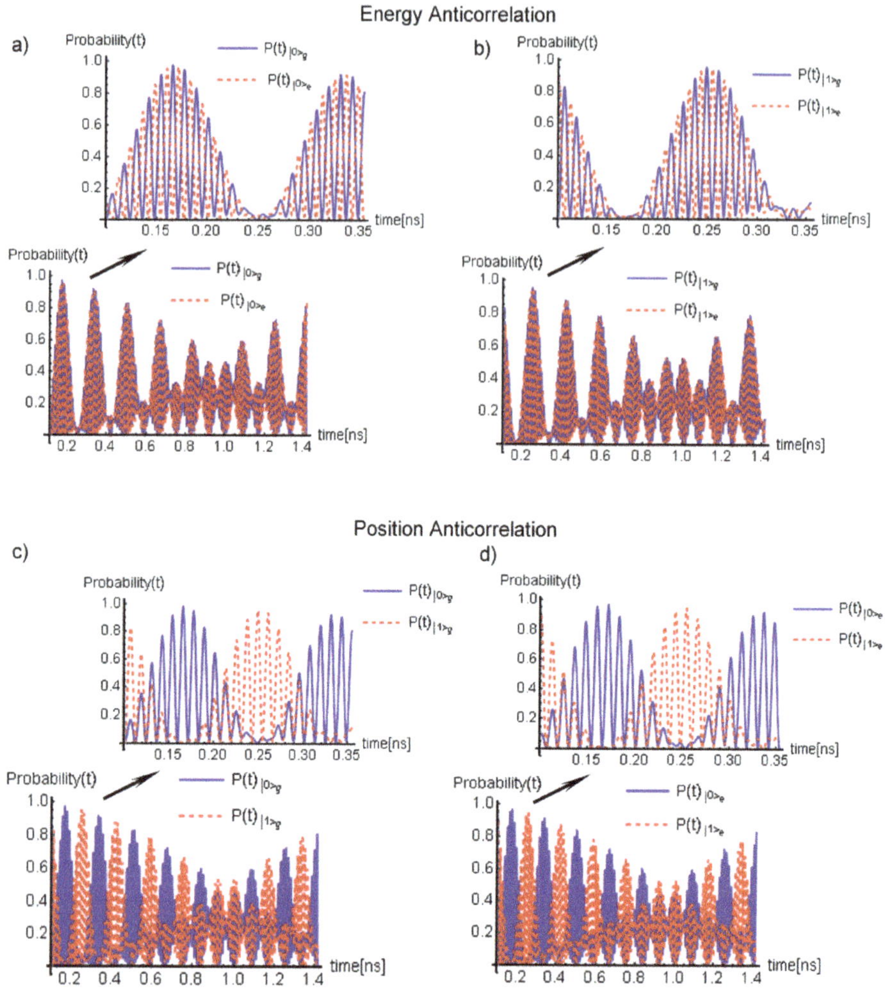

Figure 4. (**a**) Evolution of the probability of the states $|0_g\rangle$ and $|0_e\rangle$. These states correspond to the same position (localized state $|0\rangle$) but different energy levels (ground state $|g\rangle$ and excited state $|e\rangle$). It is observed that they are anti-correlated. (**b**) Time Evolution of the probability of the states $|1_g\rangle$ and $|1_e\rangle$. Similarly, these states correspond to the same position (in this case the localized state $|1\rangle$). Energy anti-correlation is also observed in this case. (**c**) The time evolution of the probability of the states $|0_g\rangle$ and $|1_g\rangle$. These states correspond to the same ground energy level (energy state $|g\rangle$) but different position (localized states $|0\rangle$ and $|1\rangle$). Therefore, they are anti-correlated. (**d**) Time evolution of the probability of the states $|0_e\rangle$ and $|1_e\rangle$. These states correspond to the same excited energy level (in this case energy state $|e\rangle$) but different position (localized states $|0\rangle$ and $|1\rangle$). They are also anti-correlated.

Finally, to conclude the analysis of the system of a single qubit, an example of a typical quantum path following the time evolution of (15) is presented in Figure 5, where a photon is absorpted during a transition from a localized ground state $|g\rangle$ to a localized excited state $|e\rangle$, while a photon is emitted during a transition from the state $|e\rangle$ to the state $|g\rangle$. After obtaining the expressions for the evolution of various amplitude probabilities from (17), we follow the numerical approach of [23] to simulate the quantum path.

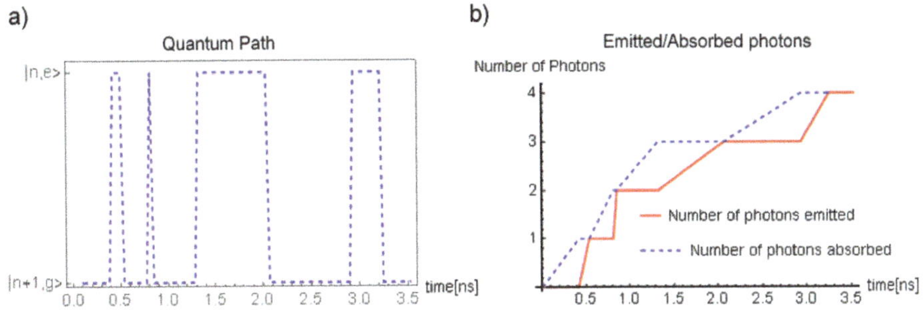

Figure 5. (**a**) Simulated quantum path as a result of transitions from the localized ground state $|g\rangle$, to the localized excited state $|e\rangle$. The transitions take place in a probabilistic manner, following the time evolution determined from the solution of (15). (**b**) Number of emitted/absorbed photons as a result of transitions between the ground $|g\rangle$ and the excited $|e\rangle$ states for the simulated quantum path. It is evident that transitions from the ground state $|g\rangle$ to the excited state $|e\rangle$ correspond to an absorption of a photon, while the transitions from the excited state $|e\rangle$ to the ground state $|g\rangle$ correspond to an emission of a photon.

4.3. Description of System of Two Entangled Coupled Position-Based Qubits with a Cavity

The system of two qubits, A and B, each one defined by a particle confined to its respective double-well, as shown in Figure 1d, can be expressed by the wavefunction:

$$|\Psi\rangle = \sum_{n_A=0_g,0_e,1_g,1_e} \sum_{n_B=0_g,0_e,1_g,1_e} c_{n_A n_B} \left| n_A^{(A)} n_B^{(B)} \right\rangle \tag{19}$$

where we again assume four quantum states for each particle. The Hamiltonian of the system can be written as:

$$\mathbf{H} = \mathbf{H}^{(B)} \otimes \mathbf{I}_4 + \mathbf{I}_4 \otimes \mathbf{H}^{(A)} \tag{20}$$

where $\mathbf{H}^{(A)}$ and $\mathbf{H}^{(B)}$ are the Hamiltonians of the first particle (system-A) and second particle (system-B), respectively, of the form (14), and $\mathbf{I}_4$ is a 4x4 identity matrix.

4.4. Maximally Entangled States and Entanglement Entropy

To investigate the entanglement and the dynamics of the system of two qubits, we will use the Von Neumann entanglement entropy S_N. The density operator is given by the expression:

$$\hat{\rho}_{AB} = |\psi\rangle\langle\psi| \tag{21}$$

Then, the Von Neumann entanglement entropy S_N is defined as follows:

$$S_\mathrm{N} = -\mathrm{tr}(\hat{\rho}_A \ln \hat{\rho}_A) = -\mathrm{tr}(\hat{\rho}_B \ln \hat{\rho}_B) \tag{22}$$

where, operators $\hat{\rho}_A$ and $\hat{\rho}_B$ are the reduced density operators, which can be found via the partial trace as

$$\begin{aligned} \hat{\rho}_A &= \left\langle 0_g^B \middle| \hat{\rho}_{AB} \middle| 0_g^B \right\rangle + \left\langle 0_e^B \middle| \hat{\rho}_{AB} \middle| 0_e^B \right\rangle + \left\langle 1_g^B \middle| \hat{\rho}_{AB} \middle| 1_g^B \right\rangle + \left\langle 1_e^B \middle| \hat{\rho}_{AB} \middle| 1_e^B \right\rangle \\ \hat{\rho}_B &= \left\langle 0_g^A \middle| \hat{\rho}_{AB} \middle| 0_g^A \right\rangle + \left\langle 0_e^A \middle| \hat{\rho}_{AB} \middle| 0_e^A \right\rangle + \left\langle 1_g^A \middle| \hat{\rho}_{AB} \middle| 1_g^A \right\rangle + \left\langle 1_e^A \middle| \hat{\rho}_{AB} \middle| 1_e^A \right\rangle \end{aligned} \tag{23}$$

Let us now assume the maximally entangled Bell state [31]:

$$|\Phi_{AB}\rangle = \frac{1}{\sqrt{2}} \left(\left| 0^A 0^B \right\rangle + \left| 1^A 1^B \right\rangle \right) \tag{24}$$

which is defined generally between any two systems A and B. Writing this state in the basis of our system with system-A (qubit #1) and system-B (qubit #2), we get:

$$
\begin{aligned}
|\Phi_{AB}\rangle = \frac{1}{\sqrt{2}}\Big(\left|0_g^A 0_g^B\right\rangle + \left|0_g^A 0_e^B\right\rangle + \left|0_e^A 0_g^B\right\rangle + \left|0_e^A 0_e^B\right\rangle + \\
+ \left|1_g^A 1_g^B\right\rangle + \left|1_g^A 1_e^B\right\rangle + \left|1_e^A 1_g^B\right\rangle + \left|1_e^A 1_e^B\right\rangle \Big)
\end{aligned} \tag{25}
$$

and

$$
\hat{\rho}_A = \mathrm{tr}_B\left(\langle\Phi_{AB}\,|\Phi_{AB}\rangle|\right) = \frac{1}{2}\mathbf{I}_4 \tag{26}
$$

In this case, from (22) $S_N = 2\ln 2$.

In general, the maximum value of S_N is $\ln N$, where N is the number of states of one qubit. The system will be maximally entangled when all the probability amplitudes c_m, for $m = 1, 2, \ldots N$, reach the same value, $c_m = 1/N$ [32].

4.5. Simulation Results

Considering the system with the Hamiltonian (20) and by solving Equation (15), we obtain the amplitudes $c_{n_A n_B}$ that describe the examined system. The evolution of the probability of the system to be found in a specific state will be given by expressions similar to (18)

$$
P_{|n_A n_B\rangle}(t) = |c_{n_A n_B}(t)|^2,\; n_{A,B} = 0_g^{A,B}, 0_e^{A,B}, 1_g^{A,B}, 1_e^{A,B} \tag{27}
$$

Then, the probability of finding each particle in a specific position and energy state will be given as a sum of combined probabilities since we assumed a combined wave-function to describe the system. In particular, we can write:

$$
\begin{aligned}
P_{|0_g^A\rangle}(t) &= |c_{0_g^A 0_g^B}(t)|^2 + |c_{0_g^A 0_e^B}(t)|^2 + |c_{0_g^A 1_g^B}(t)|^2 + |c_{0_g^A 1_e^B}(t)|^2 \\
P_{|0_e^A\rangle}(t) &= |c_{0_e^A 0_g^B}(t)|^2 + |c_{0_e^A 0_e^B}(t)|^2 + |c_{0_e^A 1_g^B}(t)|^2 + |c_{0_e^A 1_e^B}(t)|^2 \\
P_{|1_g^A\rangle}(t) &= |c_{1_g^A 0_g^B}(t)|^2 + |c_{1_g^A 0_e^B}(t)|^2 + |c_{1_g^A 1_g^B}(t)|^2 + |c_{1_g^A 1_e^B}(t)|^2 \\
P_{|1_e^A\rangle}(t) &= |c_{1_e^A 0_g^B}(t)|^2 + |c_{1_e^A 0_e^B}(t)|^2 + |c_{1_e^A 1_g^B}(t)|^2 + |c_{1_e^A 1_e^B}(t)|^2 \\
P_{|0_g^B\rangle}(t) &= |c_{0_g^A 0_g^B}(t)|^2 + |c_{0_e^A 0_g^B}(t)|^2 + |c_{1_g^A 0_g^B}(t)|^2 + |c_{1_e^A 0_g^B}(t)|^2 \\
P_{|0_e^B\rangle}(t) &= |c_{0_g^A 0_e^B}(t)|^2 + |c_{0_e^A 0_e^B}(t)|^2 + |c_{1_g^A 0_e^B}(t)|^2 + |c_{1_e^A 0_e^B}(t)|^2 \\
P_{|1_g^B\rangle}(t) &= |c_{0_g^A 1_g^B}(t)|^2 + |c_{0_e^A 1_g^B}(t)|^2 + |c_{1_g^A 1_g^B}(t)|^2 + |c_{1_e^A 1_g^B}(t)|^2 \\
P_{|1_e^B\rangle}(t) &= |c_{0_g^A 1_e^B}(t)|^2 + |c_{0_e^A 1_e^B}(t)|^2 + |c_{1_g^A 1_e^B}(t)|^2 + |c_{1_e^A 1_e^B}(t)|^2
\end{aligned}
$$

In Figure 6a, the probability as a function of time of the ground state of system-A, $P_{|g^A\rangle} = P_{|0_g^A\rangle}(t) + P_{|1_g^A\rangle}(t)$, is plotted together with that of the ground state of system-B, $P_{|g^B\rangle} = P_{|0_g^B\rangle}(t) + P_{|1_g^B\rangle}(t)$. Remarkably, the energy states of the two qubits are anti-correlated. This result is significant since the coupling between the two qubits in this model is achieved through a photon communication. For instance, in Figure 7a, typical quantum paths of system-A and system-B are plotted. It can be seen that the quantum trajectories of the two systems are anti-correlated. This is also visible from Figure 7b, where the emitted and absorbed photons between the two systems are anti-correlated as well. When system-A emits a photon, system-B absorbs a photon, and vice versa.

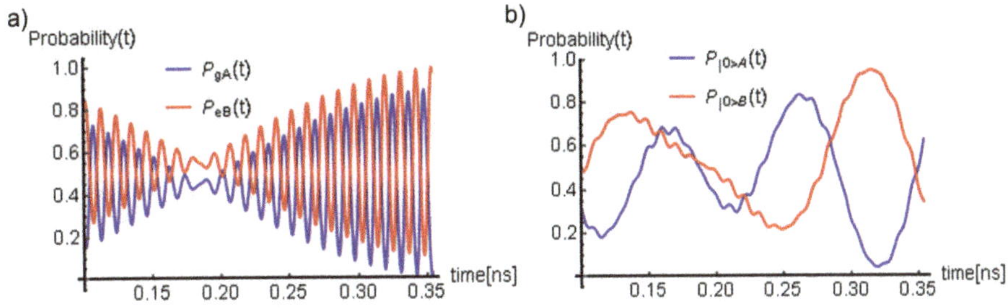

Figure 6. (**a**) Evolution of probability as a function of time of the ground state of system-A, $P_{|g^A\rangle} = P_{|0_g^A\rangle}(t) + P_{|1_g^A\rangle}(t)$, together with that of the ground state of system-B, $P_{|g^B\rangle} = P_{|0_g^B\rangle}(t) + P_{|1_g^B\rangle}(t)$. Remarkably, the energy states of the two qubits are anti-correlated. (**b**) Evolution of the probability of the localized state $|0^A\rangle$ of system-A, $P_{|0^A\rangle}(t) = P_{|0_g^A\rangle}(t) + P_{|0_e^A\rangle}(t)$, together with the evolution of probability of the localized state $|0^B\rangle$ of system-B, $P_{|0^B\rangle}(t) = P_{|0_g^B\rangle}(t) + P_{|0_e^B\rangle}(t)$ as a function of time. In this case, the anti-correlation is not trivial, and the result will depend on initial conditions. This is expected, since we assumed that there is no electrostatic interaction between the two particles, i.e., the Coulomb force is negligible. Considering this, the position of each particle is not expected to be "strongly" affected by the position of the other, only their energy states.

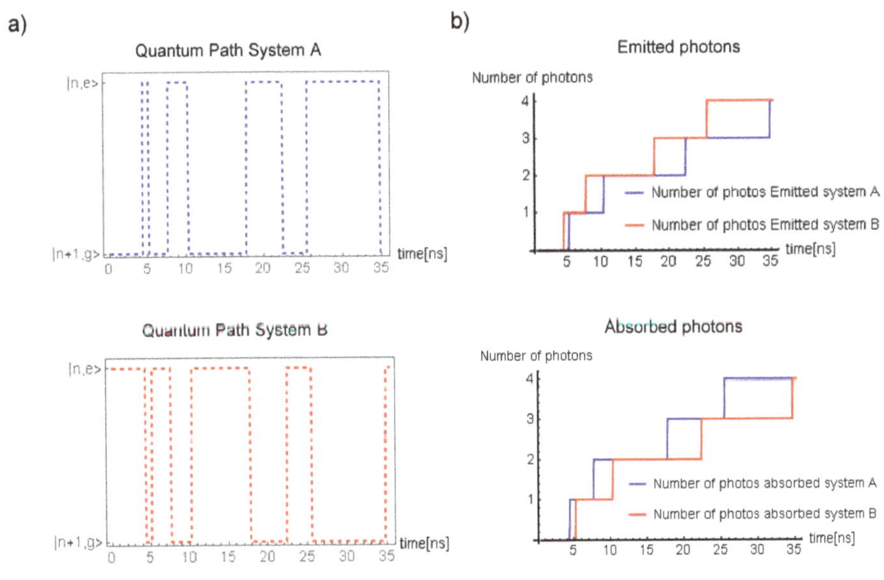

Figure 7. (**a**) Quantum paths of system-A and system-B. The methodology to obtain these graphs is similar to the one discussed for the system of a single qubit. It can be seen that the quantum trajectories of the two systems are anti-correlated. In other words, when system-A is in the ground state $|n+1, g\rangle$, system-B is in the excited state $|n, e\rangle$; (**b**) The emitted and absorbed photons between the two systems are anti-correlated. When system-A emits a photon, system-B absorbs a photon, and vice versa.

However, we also plot in Figure 6b the evolution of the probability with time of the localized state $|0^A\rangle$ of system-A, $P_{|0^A\rangle}(t) = P_{|0_g^A\rangle}(t) + P_{|0_e^A\rangle}(t)$, and the evolution of the probability with time of the localized state $|0^B\rangle$ of system-B, $P_{|0^B\rangle}(t) = P_{|0_g^B\rangle}(t) + P_{|0_e^B\rangle}(t)$. In this case, the anti-correlation is not trivial, and the result will depend on initial conditions. This is expected since we assumed that there is no electrostatic interaction between the two particles, i.e., the Coulomb force is negligible. Considering this, the position of each particle is not expected to be "strongly" affected by the position of the other, only its energy states. Here we should note that if one considers a non-negligible Coulomb force, the positional anti-correlation will be also observed. This was also demonstrated in [26] where the quantum states entanglement was due to a Coulomb interaction.

To conclude this section, we investigate the entanglement in the system under study. Figure 8 plots the Von Neumann entanglement entropy S_N defined between system-A and system-B. As shown in Section 4.4, the maximum value of S_N for this system is $2 \ln 2$. At such an instance of time, when $S_N = \max$, the quantum states are maximally entangled (inseparable). Thus, the wave-function cannot be represented as a product of the wave-functions of each particle. From the plot, one can observe that S_N is a time-dependent quantity and for this system, it reaches the maximum value of $2 \ln 2$. We conclude that the states of the two coupled qubits via the waveguide are entangled. Therefore, when a photon communication is allowed between two qubits, as constructed via the methodology discussed in this study, one is expected to achieve quantum operations between the qubits over longer distances.

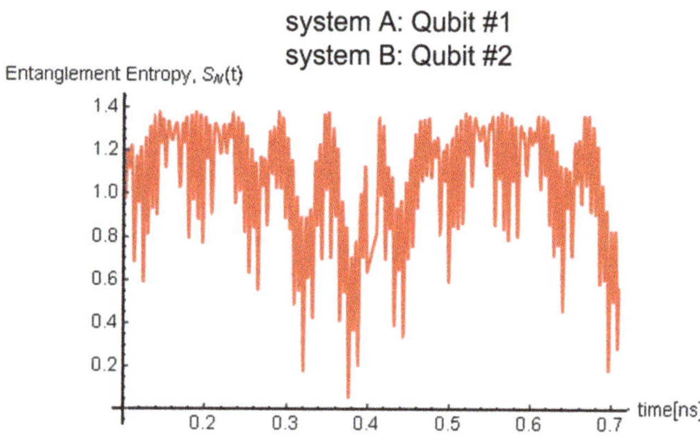

Figure 8. Von Neumann entanglement entropy S_N defined between system-A and system-B. The maximum value of S_N for this system is $2 \ln 2$. When this is the case, the quantum states are maximally entangled (inseparable). Thus, the wave-function cannot be represented as a product of the wave-functions of each particle. One can observe from the plot that S_N is a time-dependent quantity. For this system it reaches a maximum value $2 \ln 2$. The states of the two coupled qubits via the waveguide are entangled.

5. Conclusions

In this study, we investigated semiconductor position-based qubits (a.k.a. charge qubits) in the context of photon communication. Electrostatic simulations under the assumption of freezout operation show that with appropriate voltages applied to the gates of CMOS devices (based on FDSOI technology) it would be possible to achieve the desired potential energy profile meeting the requirements of a specific quantum operation. This can allow one to control the energy levels of the bound states of the system and the tunneling probabilities of particles between the barriers separating quantum wells. We discussed that the system can be approximated as quantum dots (QD), and each double QD (DQD) implements a position-based qubit.

We defined position-based qubits and their coupling with a cavity. We showed that in the presence of an external perturbative driving field, in the rotating wave approximation, the Rabi flopping frequency predicts analytically the evolution of the quantum states. The various states of the system of the simple qubit are found to be anti-correlated, both in energy-basis and position-basis representation.

We further expanded the model to include the description of the photon emission and entanglement of coupled position-based qubits based on the Jaynes–Cummings–Hubbard formalism.

We demonstrated that for a given geometry and potential energy profile it is possible to construct entangled states and trace their time evolution. We quantified the magnitude of entanglement due to the photon communication by calculating the entanglement entropy. The modelling provided in this work can offer the tools towards the optimization of relevant semiconductor photon-assisted applications and can easily be expanded in a straightforward manner to describe multi-particle systems.

Author Contributions: Conceptualization, P.G., E.B. and D.L.; Formal analysis, P.G. and E.B.; Methodology, P.G. and E.B.; Project administration, R.B.S.; Resources, R.B.S and D.L.; Software, P.G.; Supervision, E.B. and R.B.S.; Validation, P.G., E.B. and R.B.S.; Visualization, P.G.; Writing – original draft, P.G.; Writing – review & editing, P.G., E.B. and R.B.S..

Funding: This work was supported by Science Foundation Ireland under Grant 14/RP/I2921.

Acknowledgments: We would like to thank Erik Staszewski for his contribution in the preparation of the schematics. We further acknowledge Microelectronics Circuits Centre Ireland (MCCI) Dublin for Associated Project support.

Conflicts of Interest: The authors declare no conflict of interest.

References

1. Bashir, I.; Asker, M.; Cetintepe, C.; Leipold, D.; Esmailiyan, A.; Wang, H.; Siriburanon, T.; Giounanlis, P.; Blokhina, E.; Pomorski, K.; et al. A mixed-signal control core for a fully integrated semiconductor quantum computer system-on-chip. In Proceedings of the IEEE European Solid-State Circuits Conference (ESSCIRC), Krakow, Poland, 23–26 September 2019; ; pp. 125–128.
2. Leipold, D. Fully integrated quantum computing SOC in 22nm FD-SOI. In Proceedings of the International Workshop on Cryogenic Electronics for Quantum Systems, Fermilab, Chicago, IL, USA, 17–20 June 2019.
3. Bluhm, H.; Schreiber, L.R. Semiconductor Spin Qubits-A Scalable Platform for Quantum Computing? In Proceedings of the 2019 IEEE International Symposium on Circuits and Systems (ISCAS), Sapporo, Japan, 26–29 May 2019; pp. 1–5.
4. Lehmann, T. Cryogenic Support Circuits and Systems for Silicon Quantum Computers. In Proceedings of the 2019 IEEE International Symposium on Circuits and Systems (ISCAS), Sapporo, Japan, 26–29 May 2019; pp. 1–5.
5. Pakkiam, P.; Timofeev, A.; House, M.; Hogg, M.; Kobayashi, T.; Koch, M.; Rogge, S.; Simmons, M. Single-shot single-gate rf spin readout in silicon. *Phys. Rev. X* **2018**, *8*, 041032. [CrossRef]
6. Ekanayake, S.R.; Lehmann, T.; Dzurak, A.S.; Clark, R.G.; Brawley, A. Characterization of SOS-CMOS FETs at low temperatures for the design of integrated circuits for quantum bit control and readout. *IEEE Trans. Electron Devices* **2010**, *57*, 539–547. [CrossRef]
7. Bonen, S.; Alakusu, U.; Duan, Y.; Gong, M.; Dadash, M.; Lucci, L.; Daughton, D.; Adam, G.; Iordănescu, S.; Păşteanu, M.; et al. Cryogenic Characterization of 22-nm FDSOI CMOS Technology for Quantum Computing ICs. *IEEE Electron Device Lett.* **2018**, *40*, 127–130.
8. Patra, B.; Incandela, R.M.; Van Dijk, J.P.; Homulle, H.A.; Song, L.; Shahmohammadi, M.; Staszewski, R.B.; Vladimirescu, A.; Babaie, M.; Sebastiano, F.; et al. Cryo-CMOS circuits and systems for quantum computing applications. *IEEE J. Solid-State Circuits* **2018**, *53*, 309–321. [CrossRef]
9. Lutchyn, R.M.; Sau, J.D.; Sarma, S.D. Majorana fermions and a topological phase transition in semiconductor-superconductor heterostructures. *Phys. Rev. Lett.* **2010**, *105*, 077001. [CrossRef]
10. Maurand, R.; Jehl, X.; Kotekar-Patil, D.; Corna, A.; Bohuslavskyi, H.; Laviéville, R.; Hutin, L.; Barraud, S.; Vinet, M.; Sanquer, M. A CMOS silicon spin qubit. *Nat. Commun.* **2016**, *7*, 13575. [CrossRef]
11. Crippa, A.; Maurand, R.; Bourdet, L.; Kotekar-Patil, D.; Amisse, A.; Jehl, X.; Sanquer, M.; Laviéville, R.; Bohuslavskyi, H.; Hutin, L.; et al. Electrical spin driving by g-matrix modulation in spin-orbit qubits. *Phys. Rev. Lett.* **2018**, *120*, 137702. [CrossRef]
12. Corna, A.; Bourdet, L.; Maurand, R.; Crippa, A.; Kotekar-Patil, D.; Bohuslavskyi, H.; Laviéville, R.; Hutin, L.; Barraud, S.; Jehl, X.; et al. Electrically driven electron spin resonance mediated by spin–valley–orbit coupling in a silicon quantum dot. *Quantum Inf.* **2018**, *4*, 6. [CrossRef]
13. Shinkai, G.; Hayashi, T.; Ota, T.; Fujisawa, T. Correlated Coherent Oscillations in Coupled Semiconductor Charge Qubits. *Phys. Rev. Lett.* **2009**, *103*, 056802. [CrossRef]

14. Mills, A.; Zajac, D.; Gullans, M.; Schupp, F.; Hazard, T.; Petta, J.R. Shuttling a single charge across a one-dimensional array of silicon quantum dots. *Nat. Commun.* **2019**, *10*, 1063. [CrossRef]
15. Hensgens, T.; Fujita, T.; Janssen, L.; Li, X.; Van Diepen, C.; Reichl, C.; Wegscheider, W.; Sarma, S.D.; Vandersypen, L.M. Quantum simulation of a Fermi–Hubbard model using a semiconductor quantum dot array. *Nature* **2017**, *548*, 70. [CrossRef] [PubMed]
16. Van der Wiel, W.G.; De Franceschi, S.; Elzerman, J.M.; Fujisawa, T.; Tarucha, S.; Kouwenhoven, L.P. Electron transport through double quantum dots. *Rev. Mod. Phys.* **2002**, *75*, 1. [CrossRef]
17. Liu, Y.Y.; Petersson, K.; Stehlik, J.; Taylor, J.M.; Petta, J.R. Photon emission from a cavity-coupled double quantum dot. *Phys. Rev. Lett.* **2014**, *113*, 036801. [CrossRef]
18. Bose, R.; Cai, T.; Choudhury, K.R.; Solomon, G.S.; Waks, E. All-optical coherent control of vacuum Rabi oscillations. *Nat. Photonics* **2014**, *8*, 858. [CrossRef]
19. Fink, J.; Göppl, M.; Baur, M.; Bianchetti, R.; Leek, P.; Blais, A.; Wallraff, A. Climbing the Jaynes–Cummings ladder and observing its nonlinearity in a cavity QED system. *Nature* **2008**, *454*, 315. [CrossRef]
20. Akopian, N.; Lindner, N.; Poem, E.; Berlatzky, Y.; Avron, J.; Gershoni, D.; Gerardot, B.; Petroff, P. Entangled photon pairs from semiconductor quantum dots. *Phys. Rev. Lett.* **2006**, *96*, 130501. [CrossRef]
21. Deppe, F.; Mariantoni, M.; Menzel, E.; Marx, A.; Saito, S.; Kakuyanagi, K.; Tanaka, H.; Meno, T.; Semba, K.; Takayanagi, H.; et al. Two-photon probe of the Jaynes–Cummings model and controlled symmetry breaking in circuit QED. *Nat. Phys.* **2008**, *4*, 686. [CrossRef]
22. Araneda, G.; Higginbottom, D.B.; Slodička, L.; Colombe, Y.; Blatt, R. Interference of single photons emitted by entangled atoms in free space. *Phys. Rev. Lett.* **2018**, *120*, 193603. [CrossRef]
23. Lee, C.J. Photon Emission Dynamics of a Two-Level Atom in a Cavity. *arXiv* **2015**, arXiv:1512.08607.
24. Luo, Y.; Liu, N.; Li, X.; Hone, J.C.; Strauf, S. Single photon emission in WSe2 up 160 K by quantum yield control. *2D Mater.* **2019**, *6*, 035017. [CrossRef]
25. Liu, Y.x.; Miranowicz, A.; Koashi, M.; Imoto, N. Realization of symmetric sharing of entanglement in semiconductor microcrystallites coupled by a cavity field. *Phys. Rev. A* **2002**, *66*, 062309. [CrossRef]
26. Giounanlis, P.; Blokhina, E.; Pomorski, K.; Leipold, D.; Staszewski, R.B. Modeling of semiconductor electrostatic qubits realized through coupled quantum dots. *IEEE Access* **2019**, *7*, 49262–49278. [CrossRef]
27. Giounanlis, P.; Blokhina, E.; Leipold, D.; Staszewski, R.B. Occupancy oscillations and electron transfer in multiple–quantum–dot qubits and their circuit representation. In Proceedings of the 25th IEEE International Conf. on Electronics Circuits and Systems (ICECS), Bordeaux, France, 9–12 December 2018; pp. 153–156.
28. Gerry, C.; Knight, P. *Introductory Quantum Optics*; Cambridge University Press: Cambridge, UK, 2005.
29. Orlando, T. Two-Level System with Static and Dynamic Coupling; Department of Electrical Engineering and Computer Science Massachusetts Institute of Technology, 2004. Available online: https://ocw.mit.edu/courses/electrical-engineering-and-computer-science/6-728-applied-quantum-and-statistical-physics-fall-2006/study-materials/twolevel04.pdf (accessed on 20 July 2019).
30. Dubbers, D.; Stöckmann, H.J. *Quantum Physics: The Bottom-up Approach: From the Simple Two-Level System to Irreducible Representations*; Springer Science & Business Media: Berlin, Germany, 2013.
31. Terhal, B.M. Bell inequalities and the separability criterion. *Phys. Lett. A* **2000**, *271*, 319–326. [CrossRef]
32. Anaya-Contreras, J.A.; Moya-Cessa, H.M.; Zúñiga-Segundo, A. The von Neumann entropy for mixed states. *Entropy* **2019**, *21*, 49. [CrossRef]

MDPI
St. Alban-Anlage 66
4052 Basel
Switzerland
Tel. +41 61 683 77 34
Fax +41 61 302 89 18
www.mdpi.com

Applied Sciences Editorial Office
E-mail: applsci@mdpi.com
www.mdpi.com/journal/applsci

www.ingramcontent.com/pod-product-compliance
Lightning Source LLC
Chambersburg PA
CBHW062034220926
51850CB00010B/2539
* 9 7 8 3 0 3 9 3 6 3 9 8 8 *